ARTIFICIAL INTELLIGENCE

THE FUTURE OBSOLETION OF MAN

AF799921

The Basket of Fish

The Basket of Fish

ARTIFICIAL INTELLIGENCE

The Future Obsoletion of Man

The Times. They Are a-Changin'
—Bob Dylan

B. O. F. Publishing

Edited by Aundrea Fox & Scott Newman
Published by B. O. F. Publishing
ISBN Number: 978-1-7379943-8-1

Contents

The Basket of Fish

Merriam-Webster

artificial intelligence noun

1 : the capability of computer systems or algorithms to imitate intelligent human behavior

> ... helped pioneer the idea that computers can exhibit *artificial intelligence* that mirrors human thinking.
> – Richard Sandomir

> A Monmouth University poll ... found that only 9 percent of Americans believed that computers with *artificial intelligence* would do more good than harm to society.
> – Shira Ovide

> *Artificial intelligence* applications are already used behind the scenes in hospitals to automate workforce tasks, improve patient flow, for operating room scheduling and elsewhere to improve efficiency.
> – Liz Freeman

also, plural **artificial intelligences :** a computer, computer system, or set of algorithms having this capability

> Someday, our *artificial intelligences* will be capable enough to pass for human.
> – Benjamin C. Kinney

> ... they are now eager to create *artificial intelligences* that can mediate human life at every level.
> – Edward Ongweso Jr.

Assist

Artificial intelligence (AI) is the capability of computers or machines to perform tasks that typically require human intelligence, such as learning, reasoning, problem-solving, perception, and decision-making. It involves developing methods and software that enable machines to understand their environment and take actions to achieve specific goals.

Wikipedia   Encyclopedia Britannica

Auto-generated based on listed sources. May contain inaccuracies.

Ask a follow-up question   Chat

Foreword

Welcome to all. OK, let's get started.

Artificial intelligence. We've been inundated with those two words for a few years now and years before that, it was basically a concept. Not much greater than science fiction. So what exactly is artificial intelligence? To the left you'll see two definitions. The first coming from good old Merriam-Webster. Not the greatest definition mind you but a definition nonetheless. Now take a look at the one below it.

According to Firefox, this definition was created by artificial intelligence. They called it AI Assist and it appeared to me that AI Assist's definition of itself was more informative. Then it occurred to me. What the hell is AI Assist? I've seen it many times while doing regular searches so I decided to ask AI Assist a question. What is AI Assist? Below was the response.

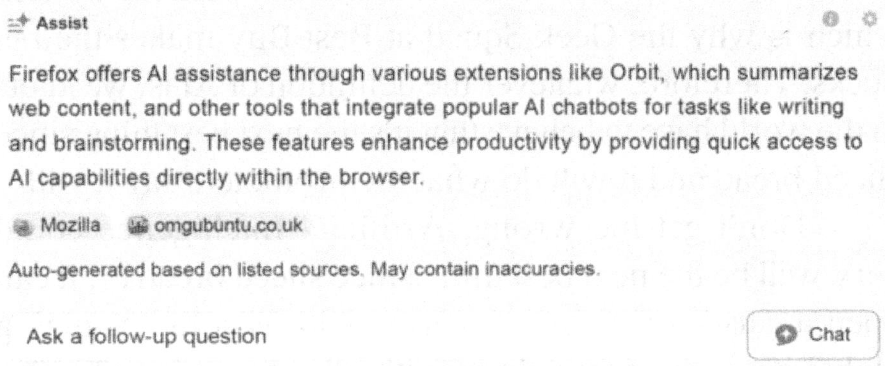

The most interesting part of both definitions from AI Assist was— did you catch it? It included the sentence, "Auto-generated based on listed sources. May contain inaccuracies." I almost found that refreshing to read because it meant that even AI is not infallible. At least right now it isn't although I'm certain the disclaimer was added for legal purposes. I wonder if there will come a day when that disclaimer won't be needed. I doubt it... pesky lawyers.

So the definition we walk away from is anything AI related is derived from some program or algorithm sitting within the confines of a computer that does something like (informationally speaking) scour the internet searching information on specific keywords then puts it all together in some coherent manner. Well, it's an educated guess on my part but it sounds right. Perhaps this is where I should add the same disclaimer. "May contain inaccuracies." I am human after all.

And that's basically it, isn't it? 99% of the worlds population aren't computer scientists. We literally have no idea how a computer works let alone its operating system which is why the Geek Squad at Best Buy makes the big bucks. Therefore, whatever the definition of AI is, we idiots in the world have to believe that it's the next best thing since sliced bread and it will do what the promoters say it will.

Don't get me wrong, Artificial Intelligence could very well be the next best thing since sliced bread or, it can open a gargantuan can of worms with vast life changing global consequences (along with all the good stuff.) The question will be, will all the promised good stuff AI will be ushering in over the next sixty-odd years outweigh the inevitable can of worms that will most assuredly require solving? More importantly, do we even know what's in that can of worms? Here's a hint. It ain't worms.

This will be the focus of this book. And we'll cover all the bases. Life in the past, life in the present and the birth of AI and the future of AI. We'll open up that can of worms and see what spills out. AI is like rolling a snowball down a hill covered in snow. It rarely gets smaller.

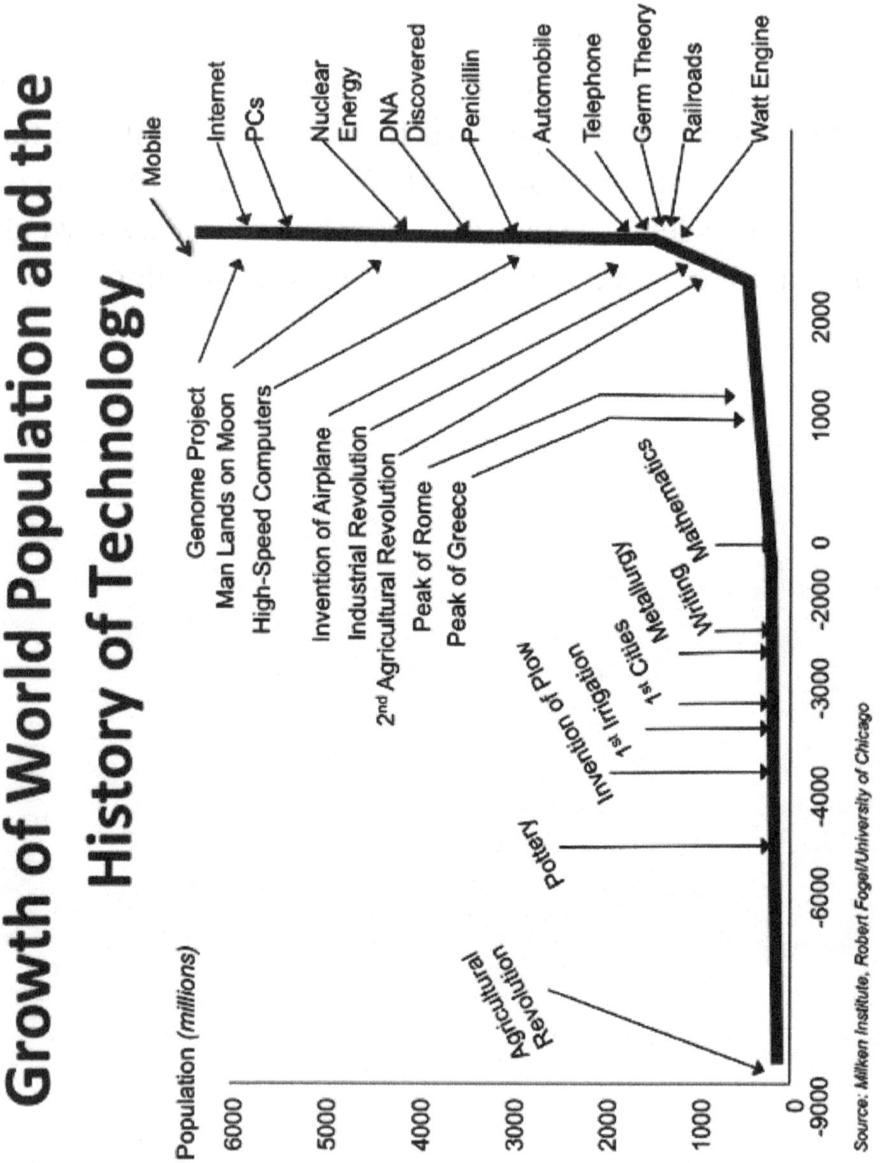
Growth of World Population and the History of Technology
Population (millions)
6000
5000
4000
3000
2000
1000
0
-9000
-6000
-4000
-3000
-2000
0
1000
2000
Agricultural Revolution
Pottery
Invention of Plow
1st Irrigation
1st Cities
Metallurgy
Writing
Mathematics
Genome Project
Man Lands on Moon
High-Speed Computers
Invention of Airplane
Industrial Revolution
2nd Agricultural Revolution
Peak of Rome
Peak of Greece
Mobile
Internet
PCs
Nuclear Energy
DNA Discovered
Penicillin
Automobile
Telephone
Germ Theory
Railroads
Watt Engine
Source: Milken Institute, Robert Fogel/University of Chicago

1
Adapt or Die

See the image on the left? Go ahead, turn the book so you can see it better, I'll wait. Interesting... huh? The chart represents progress up until the year 2000. Why just to the year 2000? First, it's the best one I could find without making a new one and second, if I had made a new one considering all the technological advances achieved in just the last twenty-five years, it would run off the page or be so small, you wouldn't be able to see it all.

Suffice to say that line on the right would continue straight up. Pretty remarkable when you consider the big picture, what we can expect in the very near future should come as no surprise. Does this include my lifetime? Partially but I can safely say that I will have shuffled off this mortal coil by the time what I sincerely believe will take place, takes place. It will be globally transformative and will eventually affect just about everyone. That is, unless you're a hunter/gatherer somewhere on the African plains or in the jungles of Brazil. You'd be safe from it all for awhile longer.

Anyone remember Walt Disney's Carousel of Progress? It was unveiled back in 1967 in Tomorrowland at the Disneyland in Anaheim, California as you guessed it, Carousel of Progress. And there it remained from 1967 until 1973. It was an enjoyable look back from the turn of the century (1900) until the 1960's illustrating the progress made in such a relatively short period of time. No matter what age you were, you kind of had to blink twice when you looked back on prior years and think about what wasn't around when you were a kid.

I'm not referring to when something was invented, I'm talking about when something became somewhat affordable to the general public. Color TV? Personal computers? Microwave ovens? Human organ transplants? Portable transistor radios? Commercial jet aircraft? The older one gets, the longer the list. Remember, it was just forty years ago when it was not uncommon to meet people (could have been your grandparents) who when asked what wasn't around when they were kids would have answered with— the automobile, the telephone, the plane and with some... even electricity itself.

We've come a really long way regarding technology in just the last forty years affecting our lives for the most part in positive ways. And when you take a look again at that chart, we accomplished this in a fraction of the time compared to progress achieved throughout prior history.

That straight line upward tells the story. And remember, that chart just went to the year 2000. Twenty-five years after that and we're into the mind-boggling age of super computer and quantum technology and it appears there's no end to it. To date, the price paid by our youth or far reaching consequences from it all is at best debatable. At least that's what everyone says but like the song says in Disney's Carousel of Progress: "It's a great big beautiful tomorrow," and to many, it will be. But unlike the progress of the past, there will indeed be a price to pay this time around. One doesn't require a crystal ball to have a pretty good idea of what's to come.

First it was fire, next came the wheel. After that, the sky was the limit just as long as it didn't interfere with church doctrine seeing that religious leaders back then ran the show and the last thing you wanted to do was piss off a high ranking member of the clergy and find your head displayed prominently on London Bridge. Not a pretty sight.

Call them inventions, discoveries, or innovation, everything new ended up displacing that which was before them. In other words, rendering them or the notion obsolete.

Wagon wheels were in use for centuries. If it was horse drawn, wagon wheels in one form or another were used. Henry Ford's Model-T in 1908 followed by the Model-A in 1928 changed all that. Unless wagon wheel manufacturers made the switch to tires, they were soon out of business or went into niche markets like western rock landscapes in Arizona.

But above all else was the invention of mass production. If an item couldn't be mass produced, nobody was interested. With mass production came lower prices and higher availability. (Just as long as there was enough supply) and created millions of jobs to satisfy demand. Best of all, in many cases, it made living one's life a little easier. That's really the point of it all, right? Progress meant making one's life a little easier to live. At least that's what one would imagine.

On the other hand, every technological revolution included drawbacks to some degree or another. Some call it the price of doing business or the price of progress while others declared under protest said price was just too damn high. With oil and chemicals came pollution. With pesticides came dead fish and birds. Science fixed some of their mistakes but not all which resulted in certain products eventually removed from the market.

Now they call it a risk/benefit analysis. Big Pharma is a perfect example. Take a look at all the commercials for one pill or another and after telling you how great it is, by the end of the commercial they've stated how you could die from it or better yet, have thoughts of suicide. Outside of that, it's a wonderful drug. Just ask your doctor.

Most everything these days comes with its own particular set of risk/benefits. Societal and environmental. Meaning, no matter what, justified or not, there will always be someone out there who will complain. It's just what some people do.

Computers will facilitate the next revolution. Don't get me wrong, they've been around for a long time and constantly improving. It's as if each new ability or feature designed into them became obsolete a few days later. And it wasn't just hardware. Thousands of programs written yearly replacing or updating those deemed dinosaurs a month later. *Old tech.* The results? Having to throw out, sell at huge discounts or donate pallets of unsold printers, motherboards and other computer components no longer desirable because of all the new and improved products thrust on the market. Think of it, a cell phone today has more computing power than the space shuttle back in the 1980's and with each passing year, we're offered a new and improved one. Just when you think you have the best, a better one comes out of nowhere. The frustrating part of it is, it sells for just a few dollars more than you paid for the one you bought the year before. You know, the last new and improved one.

Yes, home computing has advanced by leaps and bounds over just the last five years or so but that's a fraction of what computers used for industrial purposes like movies, engineering and science can do. It's like... damn! They do all that on a computer? The answer? Yes and it continues to evolve.

Computers in all their many forms have played an integral part of American life for years now. Honestly, I sincerely doubt we could live without them. We'd be back in the stone age.

But this book isn't so much about the past (though there might be a required history lesson here and there about how we got to where we are), it's about the future. Our future. The most difficult part in writing this will be keeping overlap to a minimum regarding chapters because when it comes to computers and with the advancement of artificial intelligence, it will be difficult not to but I'll do my best.

The Basket of Fish

2
The Simple Life

Ahhhhh the simple life. To some, they're living it now, to others, it's a life they aspire to live on a daily basis. But what actually is the simple life?

Take a look at the image to your left and what do you see? For myself, it represents serenity and for Norman Rockwell (one of my favorite artists) who painted it back in 1948, I believe it was his depiction of the same. It was a moment in time with the grandfather (I think, looks a little old to be the father) and his grandson on their way to spend the day fishing.

On one hand, you have a boy and his dog without a care in the world racing impatiently to do that which he's been waiting all week for. On the other hand, there's a much older man just standing, taking stock in all that's around him. Allowing his surroundings to soak in. Taking a little extra time in appreciation. To both, that is their simple life.

What's interesting about the painting is we the viewer don't look further into it other than its face value. Our curiosity goes no further than what we see other than emotions evoked by seeing two people of different generations and a dog doing one of the few enjoyable things they have in common. Fishing.

It's the feeling it represents. You could substitute just about any background or pastime and its meaning would remain the same. It could be a grandfather and his grandson on their way to a basketball game or camping by a wooded lake getting ready to try out their newly restored canoe they'd been working on for months.

There are no hidden meanings, no analysis, no questions to ponder. We don't think about what either did the days previous which brought them to this moment in time or trials and tribulations going on throughout the world. Just the joy on both faces illustrating that they are simply where they wanted to be. To us, anything more would be distracting.

For many, we spend our entire lives reaching for that goal while others appear to fall right into it with little to no real effort. Ask ten people to describe their personal *simple life* and the first thing you'll notice is that they have to really think about it first. The second will be that at least seven out of the ten will be different than the rest with two or three having absolutely no idea, though all would have no problem describing the opposite.

There was one TV show and two movies I've seen over the years that made me wonder about what my own simple life might be. The first was episode 30 of **The Twilight Zone** written and narrated by Rod Serling which first aired back in 1960 titled:

A Stop at Willoughby

Gart Williams, a tightly wound big city advertising exec suffering from an ulcer trying to cope within a job he loathes and an unsympathetic high maintenance wife Janie who thinks only of herself desperately yearns for a simpler life. Day after day he hears the yells from his boss, "Push push push Williams!" as his ulcer becomes worse.

On the verge of a nervous breakdown, the only solace afforded him is the tranquil train ride home where he can relax before dealing with his wife and her relentless complaining, who for some unknown reason, he still loves.

Pulling down the window shade to block his view of the freezing snow racing outside the train, he falls asleep. What seems like just moments later, he wakes up to find himself in an antique train car slowing to a stop.

He lifts up the shade and sees a peaceful scene of a warm sunny day with people taking a delightful afternoon stroll in a small town square with a couple of young boys dressed similarly to those in the Huckleberry Finn novels racing to the local watering hole with fishing poles at the ready. Finding it difficult to believe the incredible sight before him, he can't help but smile.

"Willoughby... this stop is Willoughby," he hears the train conductor announce walking up from behind.

"Willoughby? I've never heard of Willoughby. And look outside, it's summer. Where is Willoughby?"

"It's right outside," the conductor replies. "Willoughby... it's July of 1888. A really lovely little village. You ought to try it sometime. Peaceful, restful town where a man can slow down to a walk and live his life full measure."

He goes over to where the train cars adjoin to get a better look then feels the jolt of the train moving and wakes back up into his reality.

Concluding it was just a dream, but a dream he'd sure enjoy living in, he describes the dream to his non-responsive, uncaring wife who belittles him and the dream. Declaring she wants nothing to do with that kind of life.

The next day finds himself in the same rat-race, the same rut with the same nagging ulcer eating away at his core. That evening, back on the train for home, he closes his eyes then finds himself back in Willoughby.

"Willoughby?" he asks the conductor.
"Yes, Willoughby," the conductor replies.

He peers out the window seeing the same town with a horse-drawn buggy out front and a wave from the driver.

Gart makes the decision to exit the train but it's too late. The train jolts into motion and Gart back into reality. He swears then and there if he makes it back to Willoughby, he's getting off the train.

The following day at home and in his office, the same anguish returns once again as does the longing for Willoughby. His wish is granted on the train home when he finds himself back again in the small village which for him was the definition of peace and serenity.

He steps off the train and to his delight, he's warmly greeted by those he encounters. They all knew his name as the two boys invite him for a lazy day of fishing. Gart Williams had finally found himself exactly where he wanted to be.

The last scene of the show is back in Gart Williams reality where the train has stopped because someone had jumped from the speeding train to his death. It was Gart Williams. The irony of it all came when the ambulance arrived to pick up Gart's body. After being placed inside, the door closed with the name of the company.

Willoughby and Son Funeral Home.

Out of all the Twilight Zones, this particular episode is one of the few over the decades I remember. And interestingly enough, suicide doesn't (even to this day) enter my mind. To me, Gart Williams simply traded one disastrous, unhealthy reality for a joyful, peaceful one.

The following movie also made an impression on me in regard to a simpler life. If you've yet to see it, I highly recommend doing so. Personally, it's one of the very best Woody Allen movies ever created titled:

Midnight in Paris (2011)

Gil Pender, a successful Hollywood writer struggling with his first novel visits Paris along with his fiancé Inez who aspires to be a full-fledged member of her group of young, pompous, pedantic snobs of art and literature.

Not really Gil's cup of tea, he wanders the streets of Paris by night seeking inspiration for what he feels is missing in his story with hints of nostalgia.

After all, Paris is where all his heroes like F. Scott Fitzgerald and Ernest Hemingway did their best work with the occasional guidance of their mentor Gertrude Stein.

Gil was about to get a whole lot more than he bargained for while momentarily relaxing on a step after losing his way back to his hotel. As bells off in the distance chimed midnight, from down the street came a pristine 1920's Peugeot Landaulet. Chauffeur in the front with passengers dressed in fancy vintage 1920's clothing in the back drinking champagne. Coming to a stop in front of a bewildered Gil, a slightly inebriated man opens the passenger door of the car and with an obvious French accent, entices Gil to join them as they head to a party. "c'mon, c'mon guy— get in the car, c'mon, get in the car. We're going to a party."

Gil, a little tipsy himself agrees and squeezes in the back seat of the cab between two French girls dressed to the nines as he's handed a glass of champagne from one of them. *"The night is young. Drink up! Drink up!"*

Upon entering the large lavish party, Gil notices the piano player providing entertainment. He finds it a little odd he looks exactly like Cole Porter. An attractive young woman approaches seeing his look of dismay which only increases further when she introduced herself as Zelda. Gil mentions he's a writer working on his novel when Zelda calls out to her husband F. Scott Fitzgerald to join them. "Zelda and F. Scott Fitzgerald? c'mon, what's going on here?" Gil asks the famous couple in complete disbelief.

Gil eventually comes to grips that somehow, he's been transported back to the 1920's. He goes on to meet not just literary legends of the time but those in the art world as well like Pablo Picasso and his beautiful mistress Adriana whom he finds absolutely enchanting.

The budding relationship between Gil and Adriana is when the movie begins to reveal itself. As Gil believes the 20's was an incredible time to live in primarily because of his heroes, Adriana disagrees and declares the Belle Époque ("The Beautiful Era" between 1871 and 1914) was in fact, Paris's Golden Age and the most wonderful time of life. Ironically, this is the era right before the one Gil and Adriana find themselves in with Gil as the guest and Adriana as the permanent resident. Or is she?

As the movie moves forward with Gil's eventual meetings with Pablo Picasso, Ernest Hemingway, Salvador Dali and Gertrude Stein herself who agrees to critique his novel, it's Adriana's turn to experience the time she found so wonderful and fascinating. As with Gil's unknown invitation back to the 20's, a horse drawn carriage with a richly dressed couple in the back pulls up with the couple inviting them both to join.

The carriage ride ends in front of Maxim's, the social and culinary center of Paris during the Belle Époque. Adriana is ecstatic. While dancing, she tells Gil why she's certain the Belle Époque was an incredible time to live in with Gil beginning to see a pattern forming. From there they go to the iconic Moulin Rouge where they notice none other than Henri De Toulouse-Lautrec sitting along the stage and ask if they can join him. "Please have a seat," Lautrec replies.

During the conversation, Paul Gauguin and Edgar Degas make their entrance and join the group. Adriana is thrilled and continues her unequivocal belief the Belle Époque was the Golden Age of Paris. The movie comes full circle when Gil asks Gauguin and Degas what they thought the best era to live in was. To Adriana's surprise and disappointment, they both replied that, in their opinion, the greatest era was the Renaissance.

This is when (at least to me) the true point of the film is discovered. To Gil living in the 2000's, the 1920's was for him the best, to Adriana, living in the 1920's, it was the Belle Époque. To Gauguin and Degas living in the Belle Époque, it was the Renaissance.

In a way, the above is true when you remove all the societal ills of the times we conveniently remove from our memories as we reflect on times past with nostalgic fondness. Depending on your age, we find the older we become, we tend to say "I remember the good old days when..." or "Life was simpler back when..." with each age filling in the blanks. Interestingly enough and generally speaking, rarely is "the present" ever mentioned. Societally or culturally.

Again, could be because of selective memory by filtering out all the shit that took place. We tend to remember only the good parts. I purposely left out the ending of the movie just in case you want to see it.

Now we're in the era of computers and all that goes with it. This last movie was produced way before its time because what was described back then as a science and technological possibility is now becoming and will soon be a science and technological fact. What's the movie?

Brainstorm (1983)

The synopsis of this particular film is fairly simple however if you haven't seen it, I recommend watching it if it's still available. I find it remarkable it was produced forty-two years ago.

Basically, it's about the morality of technology. In the film, technology has advanced to where one individuals actions can be recorded and played back to another for him to see, hear and feel the same experience. Sky diving, mountain climbing, you name it. Even sex. If it can be recorded, it can be experienced by another by virtue of a neuro-link from a headset.

As usual, the film goes from using this technology for more or less entertainment until the military gets wind of the research. Seeing the real possibility for military applications, they walk through the door of the research center and assume complete control of the project under the auspice of national security.

I don't want to give away too much in case you might

want to watch it. I'll just add it's equally focused on the relationship between two of the stars: Natalie Wood (her final film) and Christopher Walken.

The applications of this new technology is staggering from fun and games to sexual encounters and let's not forget the military.

Like I said, the movie was fairly straightforward. But think about just one aspect. This will revolutionize the porn industry in fundamental ways. Prostitution will become virtually non-existent. The demand for online sex workers will greatly decrease. And believe it or not, this will also dramatically affect decreasing personal relationships. More on this later.

Seems like we're on the verge of that now with Elon Musk's Neuralink so I'm certain within the next fifteen to twenty years (perhaps sooner), this or some version of this will be readily available and affordable to the masses. Others say it will be closer to ten— fifteen. Let that sink in. The day this becomes reality, the world will become your oyster and I mean in pretty much any capacity limited only by your imagination. Moral implications? Again, use your imagination.

And this is only one small part of a very large picture but it all has to begin somewhere, right? For this particular beginning, it's Musk's Neuralink. But we can't dismiss all previous technological contributions.

3
Technology

Necessities

One could argue the beginning of modern times began with the advent of railroads and the use of oil as fuel for just about everything. Then came steel. The turn of the century with John D. Rockefeller, Cornelius Vanderbilt, Andrew Carnegie, the Wright brothers and others pretty much started it all and we wouldn't be where we are today if it weren't for them, although by today's laws and regulations, most of them would be sitting in jail for their somewhat unscrupulous deeds. But it really wasn't until the 1940's when the results from their innovation became the norm. The face of America changed.

Think about it, when it comes to the necessities of life, the human being doesn't ask for too much. A roof over our heads, decent food to eat, something to cook said food on, basic tools to fix whatever breaks, transportation, some form of entertainment and a job to earn enough to pay for it all. Everything outside of that were *bells and whistles or luxuries* which were dependent on how much one earned. Let's take a quick look at some of these necessities. Take notice that I'm saying "*when the following items were used publicly throughout most the U.S.*" not when they were invented. Two different animals.

We'll begin with... "Can you hear me now? Can ya?"

The telephone— Available generally since the 1920's
By 1945, about 45% of American households had a telephone increasing to over 90% by 1970. Believe it or not, there was a time when you could not own a telephone. It was basically leased to you when you set up your account.

While hard-wired phones still exist, for all intents and purposes, the cell phone is the leader in how we communicate. The smart-phone has greatly improved its capabilities but that didn't necessarily make it a good thing.

Automobiles— Affordable since 1925
Transportation for you and a few others from point A to point B. To this day, one hundred years later, that's still all they're built for. Sure, a few bells and whistles to justify the higher price but still not much more than transportation.

Electric Toasters— Everywhere since 1926
Outside of their look and a few improvements, toasters have worked on the same technology for the past close to hundred years. Red hot wires.

Refrigerator/Freezer— Since mid 1940's
Again, pretty much the same technology. Bells and whistles galore since, but same function. They keep food cold or frozen.

Black and White TV's— Since late 1940's
Now obsolete. Digital electronics replaced them.

Color TV's— Since mid to late 1960's
Now obsolete. Digital electronics replaced them.

Microwave ovens— Widely used since 1970
And continues pretty much unchanged to this very day. A clock, a keypad and a few presets.

Now, I could have added a thousand items to this list like toilets, indoor plumbing and the like but the point of it is, our basic living necessities haven't really changed since 1945. All upgraded to be sure but to this day, they still have the same function.

A Better Mouse-trap

Yep! Most everything we use today is basically a new and improved version of something invented decades earlier. Some cheaper through mass production, others with bells and whistles to make them more expensive. Interesting that most of the bells and whistles are electronics. We'll be getting to that a little later. I know, I know, I'm using "bells and whistles" far to often. Where's a good thesaurus when you need one?

Welcome the Age of Computers

Not since *the Industrial Age*, beginning in the mid-18th century, had such drastic fundamental changes taken place across America. However, it was the industrial age through machine-based manufacturing that necessitated the rise of factories thus allowing significant changes in transportation and communication, greatly transforming both society and the economy. In the beginning, the worker wasn't overly happy with his wages, but that all eventually changed.

The Computer Age, also known as the Information Age took the Industrial Age a thousand steps farther to an entirely new level which even today knows no bounds. The bad news was, most of it was made in China.

It all began in the mid 1960's primarily for military, government, learning institutions and big business. Rapid advancements in computer technology and digital communication transformed how we process, share, and access information but it was the personal computer (some twelve years later) that changed the world.

The personal computer revolution began in the late 1970's with the introduction of affordable microcomputers designed for individual use. I'm sure all of us old timers remember what they were. Apple II and Commodore 64 just to name a few. Sooner than later, everyone would have one in their homes.

Parents used them, kids used them for homework and games but when the internet arrived around 1983, it turned the personal computing world on its head. It was one thing sharing information but little did anyone know the impact it would have on society. And what an impact it had. Changed the whole ball game.

It gave people around the world an opportunity to meet and chat with others whom they would never have met otherwise. Relationships sparked, even marriages and unfortunately, others lost their life savings from scammers.

Today we do most tasks online from shopping to registering our cars, banking, paying taxes and purchasing just about anything under the sun, saving people time and a lot of trouble. Whatever jobs deemed no longer needed were easily replaced during this transition and opened a new market for technical support jobs primarily in India and the Philippines.

And as usual, it took less than a millisecond before porn was right up there front and center. Gone were the days of fifteen year old boys sneaking into their parents bedroom searching for Dad's stash of Playboy's for a peek.

Now all they had to worry about was someone walking in their bedrooms and be shocked at what was on little Johnny's computer screen.

With the introduction of the smart phone, kids no longer needed to hurriedly turn off their bulky computer monitors when hearing a knock at their door by Mom or Dad. Free porn exploded throughout the internet although to this day, I have absolutely no idea how they make any money at it.

Unfortunately, for young impressionable minds— it was one thing to take a peek at breasts when discovering a discarded girly magazine back in the 60's, or Dad's stash of the same in the 80's and quite another to have access to full blown hardcore porn on your phone today.

In regard to American culture, the jury is still out whether or not all this technology was a good idea. For someone like myself who spent his youth long before the personal computer, when calls were made from rotary dialed phones hardwired somewhere around the kitchen, when you spent hours talking to the girl you were sure (at fifteen) would be your future wife. When you can remember with fondness the very first show you saw on a color TV at a neighbors house because at the time, your parents were still saving up for one, when you looked forward to going to school and took pride when you showed your parents your report card and most important of all, when you could actually relate to some degree with that Norman Rockwell painting in the beginning of this chapter. I can honestly say that growing up without personal computers or the internet, I didn't miss much. In fact, upon reflection, I consider myself fortunate having grown up without it as many young people today are realizing. And smart-phones? Many are making the switch back to what are now being called dumb-phones. Meaning phones without the internet. Who knows? Perhaps the art of conversation will become popular again.

There's an old saying. Just because we're able to do something, doesn't necessarily mean we should do it and smart-phones can be a good example. Personally, I use my smart-phone (yes, I have one) for only three purposes. To make calls, to send and receive text messages and check the weather using a radar app. That's it! Then again, I'm an old fart who can barely figure the damn thing out. Lots of cool tricks smart-phones can do these days like replace a frown with a smile on a three year old's face when the fifteenth attempt didn't produce one with a couple of clicks. I saw that once in real time and I have to admit, pretty cool! But can I live without all this coolness? Absolutely. In fact, perhaps they need to change the name of the smart-phone to something like a mini-pc that makes calls because it appears making a phone call is the least important part of it.

When thinking in regard to technology, what's the first thing you think of? Wait! Don't answer quite yet. At least not without considering life itself and its necessities over the past hundred years or so.

We're still driving cars albeit now we have hybrids using fuel and electricity or just EV's. Self-driving cars will eventually become the norm although not in my lifetime. Aircraft is still the form of transportation for long trips. I doubt we'll be seeing an alternate form anytime soon. We're still purchasing our food and other supplies from grocery and/or big box retailers with most other items purchased online. We're still living in apartment complexes or single family homes with hot and cold running water, toilets and showers. So outside of new and improved, the actual necessities of life really hasn't changed that drastically over the last century.

But when in regard to the basic necessities of life, the advent of computers and the internet only changed how we procured them. Don't misunderstand, all wonderful in their own right for the sake of convenience. Some might even call them miracles. Miracles? That might be stretching it a bit. Let's take a look...

Miracles

To some it's defined in religious terms. To others, we think of medical miracles. But with all the new information coming from those who are supposed to be a whole lot smarter than the rest of us, I'm just a tad skeptical of whether these so-called medical miracles aren't doing more harm than good. Especially now that life expectancy in the U.S. has gone down according to recent studies. Big Pharma is drugging the population into oblivion. Our diets basically suck. And now there appears to be a pill that will cure all your ills. If it doesn't, I'm certain they'll come up with a new disease, illness, pandemic or some kind of syndrome they'll have a remedy available either now or shortly.

But before you call me a Debby Downer, there has been some amazing things going on in the medical industry that should be recognized. Replacement hips, elbows, knees and organ transplants are getting safer by the day. Robots are doing some intricate surgeries with success unheard of in years past and it's just the beginning.

There will be an incredible future for medicine in the form of surgery via robotics and AI. I'm convinced it won't be too long before repairing nerve damage will become routine allowing people to regain the use of their limbs. Don't get me wrong, I'm no doctor or medical researcher but I can easily see with what science and research has accomplished over the years specifically genetics and robotics, we'll in the not so distant future see personal internal organ replacement accounts. Where people can grow their own internal organs externally in a lab using a blank then covering them with cells from the recipient thus causing little to no rejection when transplanted. Again, all this and most likely much more will take place within the foreseeable future dramatically changing many lives who had no hope just a decade earlier.

The Future

Innovation within the medical community compared to the present and a relatively short period of time will look like night and day and AI will play an important role in just about all of it but the future? Too soon to talk about the future at this point in the book.

Let's just say that you thought the last hundred years were incredible due to technological advancement? It will be unrecognizable in the next hundred as you'll soon find out.

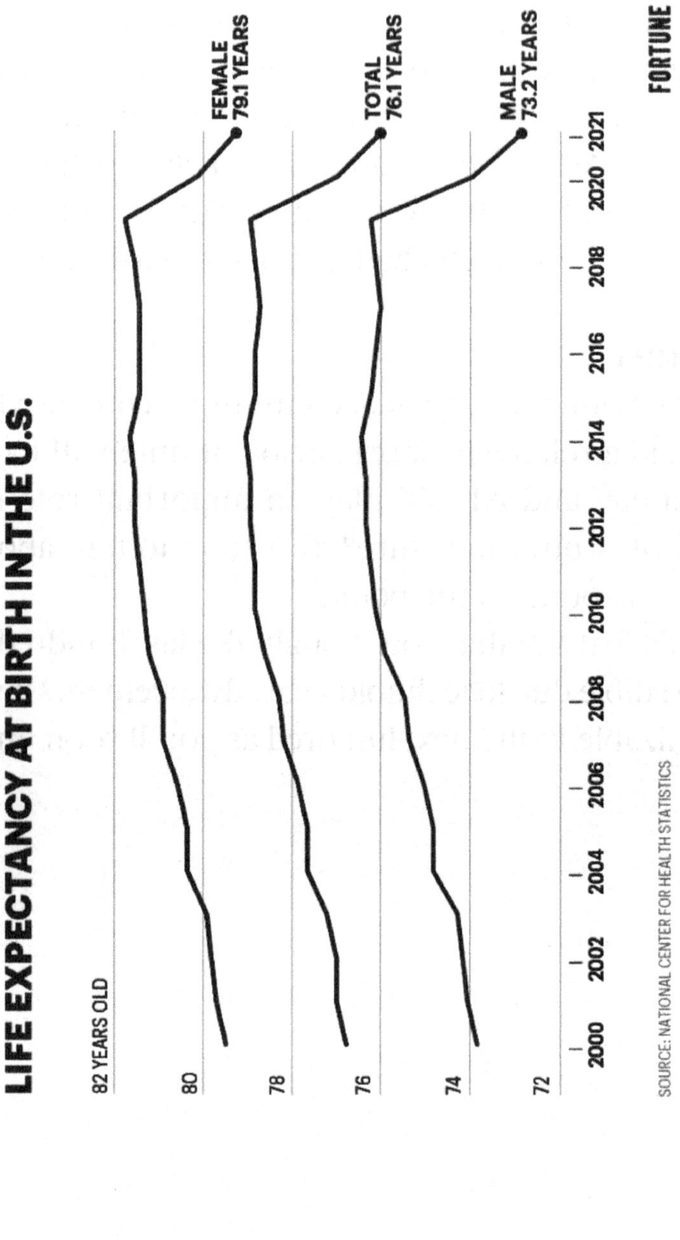
LIFE EXPECTANCY AT BIRTH IN THE U.S.
82 YEARS OLD
80
78
76
74
72
2000
2002
2004
2006
2008
2010
2012
2014
2016
2018
2020
2021
FEMALE 79.1 YEARS
TOTAL 76.1 YEARS
MALE 73.2 YEARS
SOURCE: NATIONAL CENTER FOR HEALTH STATISTICS
FORTUNE

4
Confusing Reality

This is the final chapter in Group 1 of chapters in this book. This Group was intended to give you a brief idea of how we got to where we are now. Yes, I could have gone into much more detail but if I had, this book would have turned into a $49.95 door stop that people would read only if they wanted to fall asleep.

Take a look at the chart on the left. It represents life expectancy in the U.S. from 2000 to 2021. Believe it or not, I looked up a ton of charts and this was pretty much the best I could find because all were slightly different and I couldn't find one that went up to 2025.

Some might say well c'mon, race isn't included and my response would be no matter how you look at any chart, at best there's a six to seven year difference. And when we're talking about our *golden years*— you know, those years where we're bumping into walls and peeing our pants. To me, it simply wasn't a significant difference to warrant me caring.

That being said, you can't discount that at the end of the day, that's what it's all about, right? After everything is said and done, the bottom line is how long we live and more importantly, the quality of life we enjoy while we're living it. Quality of life is key.

Right now, it's somewhere between 76 and 80 years.

A little higher for women and a little lower for blacks. We had a dip because of Covid-19 but, should go back up giving humanity a few extra years soon enough. But look at it this way, generally speaking for the majority of us, once you hit somewhere between 70 and 75— it's all downhill from there. Of course there are exceptions to every rule and perhaps you're one of them but... but... well, you get my drift.

As we move into Group 2, you're going to see how artificial intelligence is going to change that. In fact, it's going to change more than you could ever imagine.

Will we live longer? Will our quality of life remain good even with the added years or does it mean additional years of bumping into walls and wearing diapers? Right now, one can only speculate and hope our slow degradation begins just a little farther down the road of life.

The rest of the book will be completely focused on Artificial Intelligence and its future, primarily in the U.S. but one can easily apply what you're about to read to the rest of the world because if it doesn't take place congruently, I can assure you, it will very soon thereafter.

The good— the bad— and the ugly. The funny thing is that people will have no choice in the matter. They've said the same thing for the last 150 years— it's the price of progress.

5
The Rise of A I

Not in My Lifetime

The idea of AI has been around for decades. It started out as science fiction seen in comic books, films and TV going back to the 1930's. With technological advances leading up to computers, the thought of AI not only became attainable, but to some degree it already exists. Still in its infancy mind you, but the child is growing fast.

At the moment, we're in the *just look at how AI is going to be so great, so wonderful, it's going to make life so much better* phase. All good news and they won't be wrong as AI grows. Some pretty amazing things are going to happen within the next five to ten years but I personally feel it's still at least a decade away before any kind of transformative change takes hold. Subtle at first. Most won't take notice unless they're college students or working in fields AI has a foothold in.

Right now, most of us use it primarily to be amazed at what it can do in a visual way. Again, it can place smiles on children's faces in family photos after the fourteenth attempt fails. It can write term papers for students who would rather party than study. It can paint pictures, write songs, speeches and can see trends before humans figure it all out. I'm certain there is much more AI can do at this very moment— it's just not to a point where most of us care. It's like going to the circus and seeing the bearded lady. It's cool, then we go home and make dinner.

And that's really it for old farts like myself. We think about AI for a minute or so, marvel at the photo editing of grand-children, then go on about our day thinking nothing more of it.

I was talking with my daughter about writing a book about AI. We both agreed that it's here to stay but I told her I won't be experiencing the full affects of it in my lifetime. I also told her she shouldn't expect to see the real end game of it all in hers but her children, who are now in their early teens? Absolutely.

The Need for Speed

Answers in seconds for what in the past would take days of research, calculations, even thought itself because AI will be able to give suggestions, bring up its own idea of aspects we have yet to ponder. That's what we all want, isn't it? Instant gratification without having to work too hard for it.

The First Wave

The first wave has already begun and it's all visual. We're in the "WOW! Look at that! Amazing! each time we watch a documentary on AI, or a segment on the news but for the time being, most here in the U.S. have been spectators of the wizardry AI can produce. The purpose is to get people used to it. Comfortable with it even though we haven't a clue about its inner workings. AI needs to be accepted, it needs to become the norm. It needs to become a part of our everyday reality. And what better way to get hooked on AI is through visuals.

The more we hear about all the innovation in medicine and like I mentioned earlier, advancements in surgeries, transplants, prosthetics and speeding up cures, AI will increase its WOW factor exponentially. And no one can deny all this is (and will continue to be) great for the population. AI will increase the ability to save lives and add to the quality of lives of thousands around the world with much more to come. In all honesty, it is pretty amazing from what we've seen so far.

The Second Wave

The second wave is in the works even as I type but it's not quite there yet. And perhaps that's intentional. One step at a time when it affects people in a more personal way, like our ability to earn a living.

Like I said earlier, AI is all about information, speed and its use in the workplace. Remember the wagon wheel? I rest my case.

You hear on the news that AI will replace thousands of jobs human's currently fill. Guess what? It's true. Think about it. Humans are pretty damn needy and depend on others to satisfy those needs. At first it was simply food, water and shelter. Now? Our desires evolved into needs as well. Suffice to say that in order to satisfy at least 80% of our needs and desires, it takes money. Steal or earn it, it takes money. And how do we earn money? For the most part it's through our labor using either our hands, our bodies (sexually speaking) or our minds. The first jobs to become obsolete for the human being will be what we call *paper pushing* jobs. More on this very soon.

Everyone wants to save a buck and when an employer can save thousands each year and a lot of legal headaches dealing with a human in the workplace, it would be in their best interest to say *Hasta La Vista* to the human after handing him or her their final check and a coupon for a free lunch at the local bistro if AI can do the same job with relatively the same if not more efficiency.

Everything we do online are wonderful examples. We prepare/pay our taxes, register our autos, book airline tickets, hotel reservations, renew our drivers licenses, purchase goods just to name a few. It's faster, more efficient and therefore more economic, and unfortunately for some, those jobs once filled by humans no longer exist. When's the last time you saw a brick and mortar travel agency?

I talked with my brother the other night about writing this book who travels quite a lot as a keynote speaker. He told me about an instance during a recent trip where interaction with humans was basically non-existent. First, he booked, paid for and printed out his airline tickets all from the comfort of his home. When he arrived at the airport, he was not only able to confirm his flight, he checked his bags and attached his baggage claim tickets himself before handing them off to human number one to place said bags on a conveyer belt. Going through TSA was a breeze because he had already done whatever was needed to make going through TSA a breeze. But he did show his passport to a TSA agent to move along. That was human number two. After that, it was just a matter of boarding the plane. Think about that. Two people compared to what once took dozens just to board a flight.

Next was his hotel. Like his airline tickets, he found the hotel of his liking, booked the room and paid for it online. He was all set to go. But it didn't stop there. Now, believe it or not, to enter his room, he just tapped on a graphic button provided via a keyless entry app on his phone and viola, the door opened. All this without interacting with a soul. I wonder how many jobs that replaced. This my friends is just the beginning.

So, what other jobs will AI replace in the very near future? When I say very near future, I'm talking about within the next five to ten years. Yes, it's that close.

And this is the strange thing about all this. Yes, for right now, jobs are and will continue to be plentiful but the keywords are, *for right now*. So for all you young people out there in high school and college, I would seriously take a second look at how you plan on making a living because the jobs that look so plentiful at the moment might not be there. Especially the paper-pusher jobs. Stick to anything having to do with AI or the trades for those will be the jobs last to go. But let's answer the question, how many jobs will go the way of the dodo bird in the not so distant future?

The answer? Tens of thousands.

And why am I so certain about this? I'm not only certain, I'll show you how it will all take place. It's quite simple actually.

First Things First

Before you see much movement, you know— movement in the AI industry that creates headlines, those data centers we've all heard so much about have to be built and put into operation. It's already begun in some places but for AI to run at full potential, more have to be online and ready to go. Along with that is making sure the energy needs for these data center projects becomes available.

With trillions pledged from around the world, the one thing we can be certain of is the funding. Now all these industry giants will have to do is navigate what I'm sure will be mountains of legal and regulatory hoops they'll be required to jump through.

Case in point, remember the Los Angeles fires back on January 7, 2025? Burnt Pacific Palisades to the ground. Guess how many permits to rebuild have been issued since? As of June, 2025 only 54 out of some 550 applications submitted have been approved. That pretty much says it all. Now I understand not all data centers will be built in California but this gives you an idea (unless Trump clears out all the red tape) of what's possible.

So, once the data centers and energy infrastructure are in place, this is when all the real fun begins. Who knows, perhaps they won't need them all in the beginning to get things rolling. I submit the following—

We've all heard of the Amazon product called Alexa, correct? Seems like everyone has one. She monitors just about everything in the home and performs all tasks within the confines of her programming which will just increase with time as of course, AI doing the same. *"Alexa, play my favorite music. Alexa, what's the capitol of Germany?"* And so on...

Well, be prepared for the Alexa of the near future.

In this (to a large extent) semi-fictional yet extremely plausible future, I borrowed the image of an Alexa machine and used it as a way to illustrate what I sincerely believe will be a significant part of this near future. (Hope that was OK seeing as I'm using it for educational purposes.) But for this lesson, we'll call this brand new innovation NEWT. ***(Full disclosure) There is no AI product called NEWT.*** I made it up because I always liked the name. Although I will be accepting all offers from interested parties. And no, has nothing to do with Newt Gingrich for those of you who are wondering.

Over the following pages, let's take a look at what NEWT will be able to do. Shall we?

The NEWT Personal Assistant

Introducing the NEWT Personal Assistant

Powered By AI Technology

The NEWT Personal Assistant will take your business to a whole new level. For a nominal monthly subscription fee, you'll never have to be concerned with the following:

- No coffee breaks or any kind of break required.
- Not one day of paid sick leave.
- No tax reporting or withholding. In fact, no paycheck.
- No health insurance.
- No paid time off or vacation pay.
- No arriving to the office late.
- No personal baggage.
- No Human Resources.
- No excuses.

What the NEWT Personal Assistant can do:

- At your beck and call 24/7/365
- Remote communication to your personal assistant through your phone or dedicated NEWT device.
- Takes dictation and writes perfect correspondence.
- Answers calls, makes appointments.
- Makes summaries of all daily communications.
- Plenty of names and friendly voices to choose from.

- Learns what you require from a personal assistant in seconds. Can offer suggestions if asked.
- Answers questions from a wealth of online knowledge.
- Records all interactions for future reference.
- Learns through human interaction.
- Recalls all previous conversations with clients.
- Can relay messages from you to the client.
- Can have all important documents either printed or ready to print waiting for your signature.
- Consistent work performance.
- Automatic daily backups of your Personal Assistant provided for free.
- Transfers to a replacement from a defective Personal Assistant provided upon inquiry.
- Free technical support available 24/7.

Consider what the going salary is for a personal assistant these days. Forty, fifty thousand a year or more? Not so with the NEWT Personal Assistant. Let NEWT take your company to new heights.

[End description]

While a portion of what you've just read is already available, it's not completely there yet technologically speaking but trust me, it will be. And when that happens, the demand for the NEWT Personal Assistant will soar. Why? Because everyone will want one. Totally affordable although the ones saving the most will be larger businesses and corporations.

Once this transition takes full effect, the only reason you'll see a living, breathing personal assistant will be for purposes of ego. You know, someone to argue with, yell at when you want to blame someone else for your own neglect or mistakes.

Needless to say, the days of human personal assistants are numbered. And it's not just personal assistants this model can effect. A version of this model can be used for academic tutors in any subject matter. It could teach you how to do just about anything.

On a more personal note— it can be something as simple as hearing "good morning" or "good night" to a lonely individual who has no one to hear it from and sometimes, just that can make all the difference in the world— even if it's not a human being saying the words. Be honest, how often do you talk to your pet? See? (Quit smiling) But let's look into this a little deeper.

Yes, there are a whole lot of lonely individuals here in the U.S. of all ages, ethnicities and so on. Older men and women who have lost loved ones who now find themselves wondering what's in store for them. To young people finding themselves completely lost within society and no longer interested in a relationship because it's just not worth the effort. The NEWT Personal Assistant will fill that void for as long as the human feels it's necessary.

Example: There was a movie out in 2013 titled **Her** which I'm certain many did not see. If you did then you probably agreed that it wasn't that great however— the technology within the movie was amazing. The problem was, the story was horrible.

It begins with Theodore, (Joaquin Phoenix) who had recently broken up with his girlfriend. Think about it, Joaquin Phoenix in the main role should tell you just about everything you need to know about the man's demeanor. Meaning, melancholy to the point of clinical depression which we see right from the start of the flick.

Feeling rejected and depressed, he finds out about an AI operating system that includes a chat-bot who is not only able to talk with him in a personal, intimate way but does whatever task one normally does on a computer as well. He gets to choose either a male or female voice (of course he chooses female) and after answering some basic questions, his new AI friend introduces herself as Samantha (Scarlett Johansson).

Now remember, we have a very lonely, addictive, emotionally co-dependent, melancholic, depressed, dejected, rejected male having intimate conversations with an upbeat, ready to listen "female" who offers suggestions and happens to have a human personality programmed in. Having a very sexy voice, whose sole purpose is to please him is always a plus. What could go wrong.... right?

Your suspicions would be correct, he falls in love with his delusion of her and here's where the story becomes a little whacky. She falls in love with him as well because she's constantly learning through interacting with him and every article on the nature of humans available on the net which for her, can be absorbed in seconds. By this time, anyone left watching this movie comes to realize nothing good is going to result from this relationship between this particular man and a constantly evolving AI woman. It makes one want to turn the movie off because it's becoming not only predictable but damn depressing to continue watching and it's not just the characters, it's all the surroundings as well. Minimalist box shaped rooms with equally depressing people around him. And if things couldn't get any worse, Samantha begins contemplating her own existence.

Theodore's growing insecurity forces him to ask questions that Samantha knows will ruin their relationship when answered. By virtue that she's AI and can do multiple tasks at once, Theodore asks Samantha if she's talking to other men. She replies in the affirmative then as expected, he immediately asks *"How many?" "A little over six thousand"* she replies.

As Theodore looks as if he's long overdue for a nervous breakdown, he asks the two questions that when answered will become the straw that breaks the camel's back. *First, is she in love with any of them?* and Second, if so, *how many*? The first answer is *yes* and the second answer is *a tad over six-hundred.*

Right then and there, you expect his head to explode but he just sits on a staircase in public with people walking by taking absolutely no notice of him in silence. What destroys the entire concept of this film is how Samantha attempts to explain herself using what most would consider your typical girl comment which is... *"I'm just so confused!"*

Needless to say and again, as expected, Samantha informs Theodore that she needs time to figure it all out. Fortunately for her, she has an AI mentor perceived as male (of course) introduced to Theodore earlier as Alan Watts (Brian Cox.) No chance of jealousy there, right? Forgive me, have to stop laughing.

The ending pretty much goes full circle and sends Theodore back to where he started. A lonely, emotionally co-dependent, melancholic, depressed, dejected and once again rejected— male.

Could this ever happen? Can someone literally fall emotionally in love with a voice in a can or on your computer? The answer is of course they can.

Here are a couple of headlines from Japan Insider back in 2022: ***Japanese Men Are Falling in Love with Characters in New A.I. Dating Game*** and the second: ***AI Girlfriends: Exploring the World of Virtual Companions.*** Then another from Live Science published July 14, 2024 titled: ***People are falling in love with A.I. Should we worry?***

All I can say is if all this is becoming a real concern, what does that say about the world and society we live in? When we're able to receive emotional support from that which is programmed to give emotional support, will that be enough? Keep in mind, AI ten years from now will look nothing like AI does today. It will become more real and humans (especially men) are going to suck it up like a sponge. And do you believe it's going to stop here with AI in a can or on a computer? Nope, this is just the beginning until new technology evolves. And it will.

The concern is that intimate relationships between people will turn into friendships. People will end up preferring ***the friend zone*** because anything more according to them is becoming too complicated. Too many strings. Too many expectations and here's a question you're probably never asked. If one takes sex off the table in regard to relationships, would that affect your choice in who you'd want? Or whether or not you'd want a relationship at all? Good question... huh? Well, what that leaves is the NEWT Personal Assistant which pretty much takes care of most other needs and desires. Basically, it's phone sex in a can. Is it the easy way out? Damn right it is. Is it a good thing? That depends on a variety of variables but I will say it will be good for the elderly. For those younger men and women out there, primarily between 20 and 40, I suggest you get your shit together. You're going to have to come up with your own solutions.

What's next in the realm of AI? Go ahead and turn the page. If this isn't interesting enough, trust me, It will be.

The NEWT Receptionist

Introducing the NEWT Receptionist

Powered By AI Technology

Welcome to The NEWT Receptionist. The perfect solution for businesses large and small, just starting out or established. For a nominal monthly subscription fee you'll never have to be concerned with the following:

- No coffee breaks or any kind of break required.
- Not one day of paid sick leave.
- No tax reporting or withholding. In fact, no paycheck.
- No health insurance.
- No paid time off or vacation pay.
- No arriving to the office late.
- No personal baggage.
- No Human Resources.
- No excuses.

What the NEWT Receptionist can do:

- At your beck and call 24/7/365
- Remote communication to your personal assistant through your phone or dedicated NEWT device.
- Takes dictation and writes perfect correspondence.
- Answers calls, makes appointments/reservations.
- Makes summaries of all daily communications.

- Plenty of names and friendly voices to choose from.
- Learns what you require from a receptionist in seconds. Can offer suggestions if asked.
- Can be tailored to your specific business.
- Records all interactions for future reference.
- Learns through human interaction.
- Recalls all previous conversations with clients.
- Can relay messages from you to the client.
- Can have all important documents/invoices printed or ready to print waiting for your signature.
- Consistent work performance.
- Automatic daily backups of your NEWT Receptionist provided for free.
- Transfers to a replacement from a defective Receptionist provided upon inquiry.
- Free technical support available 24/7.

In essence, the NEWT Receptionist is similar to the NEWT Personal Assistant but with greater capabilities and less on personal aspects. Can be used in most any businesses reception center. With its built in cam, it can even recognize previous customers. Very little training required.

[End description]

Between the NEWT Personal Assistant and the Receptionist, you have all bases covered. Personally and dealing with the general public. Let's briefly consider these first two products and what their obvious ramifications via their wide range of applications will be. Here are just a few off the top of my head. In no particular order.

Walmart greeter— Doctors office— Academic tutors— Dental office— Automotive front office— School front office— Secretaries— Restaurant/Hotel reservation desk, fast food and pretty much any other business that currently requires humans close to the front door to greet, answer basic questions or point a customer in the right direction or notify a company employee the customer/client has arrived for their appointment.

All those jobs currently held by humans will no longer exist and we're talking hundreds of thousands of jobs— gone. Especially when AI reaches the point when it will be audibly indistinguishable from humans. We're talking hundreds of billions of dollars saved. Not just the U.S. but the world. We saw the beginning of this years ago but at the time it was limited to phone calls. Welcome to so and so, for this department, press one, for another department, press two and so on. To this day it's been a royal pain in the ass to deal with. That will all dramatically change once AI takes over.

The next model of NEWT you'll find on the following page is far more intricate. Thus the subscription fee will be significantly greater but even with the higher fee, those who utilize this model will end up saving their business far more when compared to humans. In money and efficiency.

NEWT Radiology Diagnostic Assistant

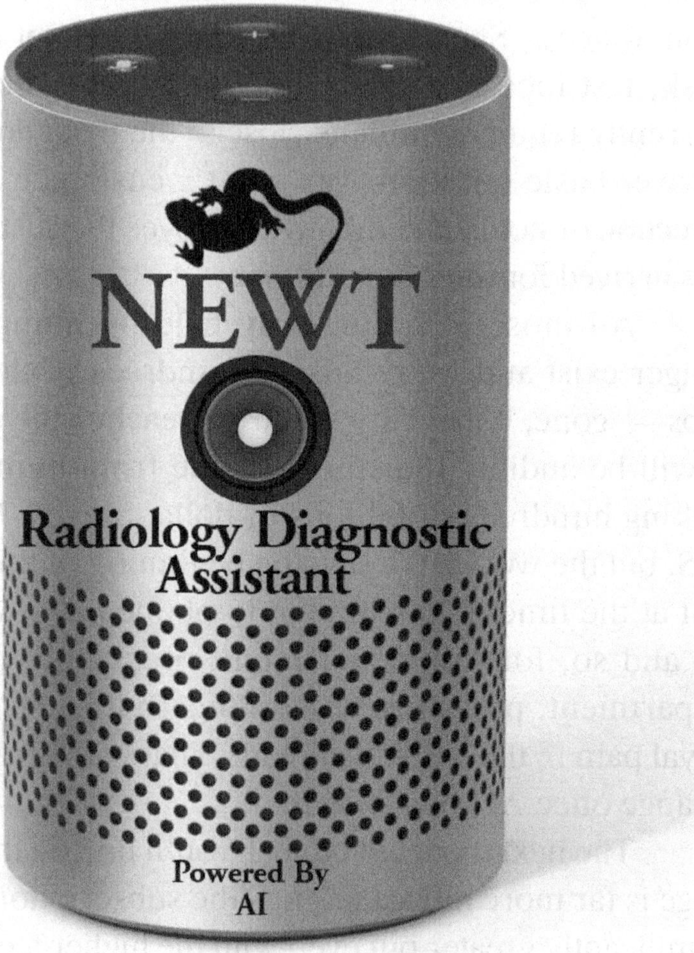

Introducing the NEWT Radiology Diagnostic Assistant

Powered By AI Technology

The NEWT Radiology Diagnostic Assistant is in a class of it's own. For a nominal monthly subscription fee you'll never have to be concerned with the following:

- No coffee breaks or any kind of break required.
- Not one day of paid sick leave.
- No tax reporting or withholding. In fact, no paycheck.
- No health insurance.
- No paid time off or vacation pay.
- No arriving to the office late.
- No personal baggage.
- No Human Resources
- No excuses.

What the NEWT Radiology Assistant can do:

- Performing and interpreting diagnostic procedures such as X-rays, CT scans, MRIs, and ultrasounds.
- Collaborating with physicians to evaluate patient health and determine appropriate radiologic techniques.
- Provides diagnosis and treatment recommendations based on the interpretation of imaging results.

- Maintaining and updating patient records with the results of imaging procedures.
- Participating in multi-disciplinary meetings to discuss patient diagnosis and treatment plans.
- Updated with the latest advancements in radiology and participating in continuous professional development
- Provide expert advice on medical imaging techniques and procedures.
- Recommend further diagnostic tests or treatment options.
- Can interface with most radiology equipment.
- Constant updates with all available data pertaining to Radiology.

Additional tasks—

- At your beck and call 24/7/365
- Remote communication to your Radiology Assistant through your phone or dedicated NEWT device.
- Takes dictation and writes perfect correspondence.
- Answers calls, makes appointments.
- Makes summaries of all daily communications.
- Plenty of names and friendly voices to choose from.
- Learns what you require from a Radiology Assistant in seconds. Can give suggestions if asked.
- Answers questions from a wealth of online knowledge.
- Records all interactions for future reference.
- Learns through human interaction.
- Recalls all previous conversations with clients.

- Can relay messages from you to the client.
- Can have all important documents either printed or ready to print waiting for your signature.
- Consistent work performance.
- Automatic daily backups of your Radiology Assistant provided for free.
- Transfers to a replacement from a defective Radiology Assistant provided upon inquiry.
- Free technical support available 24/7.

The average salary for a diagnostic radiologist in 2025 ranges from $400,000 to $500,000 per year, depending on experience and location. Around $206,920 for entry-level.

With NEWT, you get an experienced Pro right out of the box. Time, money and efficiency will take your company to new heights and can potentially save millions in the process.

[End description]

Think an item like this wouldn't sell like hot cakes? Think again. To have the ability to consult with an entity that has the ability to access the worlds knowledge on a particular topic? Let's just say whoever comes up with an AI model for this will make a fortune.

The same goes for our final product which will also fundamentally change the following profession.

The NEWT Paralegal

Introducing the NEWT Paralegal

Powered By AI Technology

The NEWT Paralegal is unlike any Paralegal you've ever worked with. This top of the line AI paralegal is all you'll require to get the job accomplished in no time flat. For a nominal monthly subscription fee you'll never have to be concerned with the following:

- No coffee breaks or any kind of break required.
- Not one day of paid sick leave.
- No tax reporting or withholding. In fact, no paycheck.
- No health insurance.
- No paid time off or vacation pay.
- No arriving to the office late.
- No personal baggage.
- No Human Resources.
- No excuses.

What the NEWT Paralegal can do:

- Conduct legal research.
- Draft legal documents: Including contracts and affidavits. Printed and ready for signatures.
- Conduct interviews, gather client information. Maintain communication to ensure effective case preparation.

- Organizing and maintaining case files.
- Assisting with trial preparation.
- Filing documents with courts via an online service.

Additional tasks—

- At your beck and call 24/7/365
- Remote communication to your Paralegal through your phone or dedicated NEWT device.
- Takes dictation and writes perfect correspondence.
- Answers calls, makes appointments.
- Makes summaries of all daily communications.
- Plenty of names and friendly voices to choose from.
- Learns what you require from a Paralegal in seconds. Can offer suggestions if asked.
- Answers questions from a wealth of online knowledge.
- Records all interactions for future reference.
- Learns through human interaction.
- Recalls all previous conversations with clients.
- Can relay messages from you to the client.
- Can have all important documents either printed or ready to print waiting for your signature.
- Consistent work performance.
- Automatic daily backups of your Paralegal provided for free.
- Transfers to a replacement from a defective Paralegal provided upon inquiry.
- Free technical support available 24/7.

What's the going salary for a human paralegal these days? Somewhere between sixty to seventy-five thousand per year. Not so with the NEWT Paralegal. Time, money and efficiency will take your company to new heights.

[End description]

Taking your company to new heights will be the understatement of the year when the NEWT Paralegal arrives on the market. There will be a model for every aspect of law from contract law to immigration law. You name it, the mythical NEWT Paralegal will easily handle it. This paralegal will have every ounce of legal data available. From past case law to legal decisions including precedent setting cases.

From legal fees and retainer estimates to whether or not you should take a case to begin with. This future technology will end the role of a paralegal. I'd even go as far to say that eventually, human lawyers themselves, their jobs could eventually be on the chopping block.

This brings a whole new meaning to the old song, ***"Anything you can do I can do better,"*** set originally between men and women.

And do you honestly believe this will stop at lawyers alone? Hell no. Judges themselves will be next. All it takes is looking at partisan judges and their rulings with many being overturned by a higher court. With AI, this would never take place. AI would be instrumental in the legalities of most any question brought forth.

Within ten years, the four products tailored to specific needs will revolutionize jobs previously held by humans and everyone will love it except the humans whose jobs no longer exist. Remember what I said earlier, this is the price of progress. And keep in mind, this is only the initial bill. Make no mistake, the products mentioned so far including what AI is doing now will take a pretty large toll on American jobs which the issue only now being seriously discussed. Personally, these people are missing the larger picture because well, what's the old saying? ***You ain't seen nothing yet.***

Described in the next chapter—

This is when the First of Four Paradigm Shifts Begins

6
AI and the Rise of Robots

Robots seem to have been around forever but for this book, we'll stick to computer controlled robots which began in the 1960's but took off like a bat out of hell ten years later. These days, robots play a substantial role in manufacturing, assembling, sorting, packaging, storage and moving of countless goods we purchase on a regular basis. Let's just say you'd be hard pressed to name anything manufactured that some form of robot didn't play a role in. Large or small.

First time we really heard about robots was in the auto industry. Sparks flying as huge robot arms did their welding. Next were computer electronics soldering micro-circuits. Now they have robots to kick out bruised or bad fruit and vegetables on a conveyer belt before they reach packaging. Even perform delicate surgeries on humans.

Think of all the jobs lost over the years. But so far, so good. From what I understand, still plenty of jobs available. That is, if you have enough intelligence to do the job in question. And what jobs are they? Unskilled labor? Those will be falling by the wayside soon enough.

For the time being, there are still jobs that require autonomy. In other words— jobs that require a human to perform. Take a moment and think about that. What jobs available right now can only a human perform? You think and I'll take a break.

So, what did you come up with? Amazon delivery drivers? Electricians? You know, everything that has to do with the trades, right? We see Mike Rowe on TV all the time saying there are virtually millions of jobs out there for the trades and right now, he's absolutely correct. (Keywords being *right now.*) OK, what else? Car mechanic? Home builder? Carpenter? Roofer? Truckers? I'm sure there are plenty of others I might have missed but I think you get my point. Plenty of jobs.

And what do these jobs have in common? Again, for the most part, they all require a human (or reasonable facsimile) to do the actual work. Automation has taken over most mass manufacturing to the point where the only humans you see in warehouses are a few people to make sure the automation ran smoothly or in the programing of said automation.

One of the best examples of this began in the late 1970's when container shipping companies began using partial automation in regard to standardized containers and automated cranes costing thousands of jobs.

Jump ahead almost fifty years and the automation of today is just about complete to the point dockworkers are striking in some ports out of fear they'll be next to lose their jobs to automation. Why? Because shipping companies no longer see the need for most dockworkers with many ports already making the switch to full automation. And while for the time being, some dockworkers may keep their jobs, unfortunately, they're just delaying the inevitable. Chalk it up to the price of progress, right? And this is just a very small part of what's to come.

Welcome Optimus

Elon Musk is developing a humanoid robot called Optimus. For those of you who have never heard of this, I invite you to look it up because it will be this technology that will replace all those trade jobs mentioned above.

There was a movie out in 2004 (damn, 21 years ago) titled **I, Robot**. While the plot was interesting, I couldn't help noticing all the robots. Seemed as if there was one for every person and one for every job imaginable. I wondered to myself, with all those walking, talking robots, what did all those humans do for a living? Had to be something for them to afford a robot walking behind them carrying their shopping, right? But, I honestly couldn't think of an answer and the movie itself really didn't supply one.

According to Musk's predictions, Optimus will be ready for market sometime in 2026. Personally, I think it's a tad overly ambitious roll-out date but sometime between now and 2030? Much more likely. And while Musk says his new robot will be affordable, (you have to ask, affordable to who?) it will be an additional five years or so before we see something similar to a Henry Ford Model T release. The day that happens will be the day the U.S. enters **Paradigm Shift Number One.** But it's not what you think. Musk gives the impression that his robots will be for general public use.

Robots everywhere! Walking our dogs, taking out the trash, grocery shopping, taking kids to the park and piano lessons. You name just about any task and there will be an autonomous robot that will do it efficiently and most importantly, without complaint.

Robots will most likely be sold in dealerships. Or perhaps leased like the telephone was years ago by Ma Bell. Remember, for a long time, the public weren't allowed to own their own telephones.

Because of the maintenance, cleaning and tuning up including dealing with all the bugs which are sure to happen on the initial rollout, leasing just might be the way to go. What might this all look like? But wait just a second there pard. Why? Because that's not how it's going to happen. Oh, it will eventually happen. You too could be the proud owner of a robot but not right away.

Musk's robots will be designed for the workforce in industry first and there are many reasons why. Here are just a few. First— it's a whole lot simpler to program an autonomous robot to perform specific, repetitive tasks as it would to clean your house or do your grocery shopping. Second— sales for robots used for industry and large companies compared to personal use will be 500 to 1 and I'm being generous. Third— the cost to manufacture robots for use mentioned above will be astronomically cheaper. On top of which. No showroom floors, no massive promotions, just volume sales from a CEO after a boardroom pitch from a few guys in suits promising lunch afterwards. This my friends is how the roll-out of autonomous robots will take place.

Remember that can of worms we discussed earlier? We're about to get to that but first...

Of course, there will be some promotion for personal robots but not right away. You'll begin seeing them here and there produced by smaller companies for sale to the egocentric wealthy to show off at parties. Which is what took place with automobiles before Henry Ford jumped into the game. Super expensive but amazing nonetheless. And that's how it all begins. It creates demand.

By this time, our imaginary company NEWT has been in the game quite awhile with it's personal and Professional line of NEWT Assistants and has the financial resources to blow away the competition after declining Musk's offer to purchase the company for twice it's value.

NEWT has garnered 68% of all autonomous robot sales in the world and now they've brought AI to the level where Robots for personal use are ready and available. Dealerships built, call centers at the ready to answer any and all questions. Let the promotions begin!

The NEWT Model - T

Introducing the NEWT MODEL - T

Powered By AI Technology

The NEWT Model - T personal android will fundamentally change your life. For a nominal monthly subscription fee you'll never have to be concerned with the following:

- No coffee breaks or any kind of break required.
- Not one day of paid sick leave.
- No tax reporting or withholding. In fact, no paycheck.
- No paid time off or vacation pay.
- No personal baggage.
- No excuses.

What the NEWT Model - T can do:

- At your beck and call 24/7/365
- Remote communication to your NEWT Model - T through your phone or dedicated NEWT device.
- Cooks, cleans, assists, can accomplish all basic household chores.
- Answers calls, makes appointments.
- Plenty of names and friendly voices to choose from.
- Learns what you require daily in seconds. Can offer suggestions if asked.
- Records all interactions for future reference.

- Learns through human interaction.
- Recalls all previous conversations.
- Can relay messages from you to almost anyone.
- Can have all important documents either printed or ready to print waiting for your signature.
- NEWT Personal Assistant capabilities integrated in base model.
- NEWT "Handybot mods" available for professional home maintenance and repair.
- NEWT "ProMods" available for clients who desire both personal and professional NEWTS.
- Subscription includes tuning and repair if warranted.
- Automatic daily backups provided for free.
- Transfers to a newer model provided upon inquiry.
- And much much more...

[End description]

What you won't see on the brochure is (if I may be blunt) what you're getting is a slave who will never have the idea that it's a slave. Will that be offensive to some? Count on it. It all depends on how one defines the term in this particular application. And the more robots look like humans, the more we'll see the term slave taking on a view similar to views we had before 1865. Additionally, it depends on the degree of personality robots like NEWT are given, though the saving grace to this issue would be to make sure all robots continue to look like robots but even then, that visual can be overcome over time. Again, it all depends on how "human" robots are perceived to be through their interactions.

In the near future, especially as the price of Robots decrease, I doubt there will be a job left that a robot won't be capable of doing. Just two key functions have to be addressed in order to complete this first paradigm shift. (More on this and the three other shifts later on.) The first is manual dexterity. It's all in the hands, right? When robots are able to screw in a lightbulb or replace a front doorknob, that resolves the manual dexterity issue.

Second is anticipation. The ability to anticipate an outcome be it probable or improbable such as noticing a glass bowl getting too close to a table's edge and moving it. Or some pet sniffing around where it shouldn't be or a child finding a knife, firearm or trying to open something that could hurt or even kill them if successful. Some may call this attribute insight. To me, it's all semantics. Call it anticipation or insight, when that day comes AI has now become as close as AI can in this regard. This is when the First Paradigm Shift is achieved.

That being said, will everyone and their brother run out and get one? That's a tough question. Don't get me wrong, millions will. Families with kids will be first in line. Next will be grandparents looking for one for when visitors come calling or setting up BINGO night at the community center but when you really give it some thought, it begs the question— outside of the novelty, would you yourself shell out thousands for one? I'm certain the average single adult or young married couple would say no, leaving the Professional NEWT Model - T remaining. It's this market Robots powered by A.I. will find its forever home. They'll fly out of dealerships so fast, manufacturers won't know what hit them. They'll find themselves hard-pressed not only to keep up with production, the most difficult part of it all will be the subscription service, on-site repair and on-site tech support. I can imagine how thick the owners manual might be though it'll most likely be online now that I think about it, most issues can be solved or explained by the robot itself in the course of interacting with its new owner. If I had to guess, I'd say The NEWT Model - T Professional will make up at least 63% of total subscription sales, if not more. Nothing better than a robot not only paying for itself but making billions in profits for corporations in the process.

But we can't discount the importance of the personal version of The NEWT Model - T for future owners. They won't be solely looking for their lawn mowed or pool cleaned. Consider the elderly just looking for conversation while the Model - T cooks their dinner or gently wakes them in the morning. Many just wanting not to be alone any longer. Someone to talk with, you know? I wonder if you caught that..... I said "someone to talk with." You see? This is how it all begins. It starts with an "Oops!"

Can the NEWT Personal Robot really change ones life to a measurable degree? Well, think about this next segment and really let it sink in.

The advantages of having your own personal robot will be much more than just cleaning your house, grocery shopping or walking your dog.

Matrix Technology

For those of you who have never seen the movie **The Matrix**, in one scene Neo (Keanu Reeves) and Trinity (Carrie-Anne Moss) find themselves on a rooftop next to a B-212 helicopter. After fighting off agents, Neo asks Trinity *"Can you fly that thing?"* Trinity replies, *"Not yet,"* then proceeds to call ***The Operator*** (one of the team played by Marcus Chong) and asks for the pilot program for the B-212 helicopter. Seconds later after uploading the program into her mind, Trinity is now a B-212 helicopter pilot.

Science fiction? Nope, not any longer. This will be a standard feature with the NEWT Model - T. Below are potentially, what you're never have to do again (in no particular order).

- No need to call an electrician.
- No need for a lawn service.
- No need for a plumber.
- No need for childcare.
- No need for an A/C technician.
- No need for auto repair.
- No need for a designated driver.
- No need for a taxi.

- No need to look for someone to help move a couch.
- No need for a house painter.
- No need to repair a hole in your drywall.
- No need for a computer repair man.

That's a hell of a lot of *"no needs"* and that's just off the top of my head. All those *"no needs"* represent three things. First is, it will save you thousands each year. To some families, it could mean a savings in the tens of thousands. Second will be time and headaches trying to find a human who will be on time and do the job properly and unfortunately, third is that all those *"no needs"* represent the eventual loss of jobs humans currently fill. How many jobs? I'd guess well over one million. So much for all those trade jobs, right?

Why call an expensive plumber or electrician and wait hours or even days for them to show up when you can ask your NEWT Model T to download from NEWT Services (for only $14.95) everything it requires to repair a leaky faucet or replace a broken electrical outlet in seconds?

I know what you're thinking.... WOW! Right? Is all that possible? Not only possible, but probable. Hence the paradigm shift and as interesting as that was, **Paradigm Shift Number Two** will take place not too long afterwards. And trust me, you thought **Paradigm Shift One** was big? No— small potatoes compared to what you're about to read on the following pages. NO PEEKING! I know, I know... you peeked.

You see, the First Paradigm Shift was primarily

about work. This upcoming Second Paradigm Shift will affect everyone in a much more personal way and by extension, society itself.

Robots like those described on the previous pages are what might be considered semi-personal by virtue of the kind of interaction taking place with humans, this next NEWT Robot will be personal to intimate. Why? Because that will be this robots prime directive. Happiness through companionship. Personal, physical and sexual. Keyword being sexual. I'm certain we've all come across a documentary or perhaps a B-movie on the topic of sex dolls at one time or another. Right now, the technology (and appearance) is still a little on the cringy side although from what I've seen, their looks have greatly increased appearing more real by the year.

Now imagine, just imagine if the (again, fictitious) NEWT Corporation developed an android that would be almost indistinguishable from the real deal? Skin, hair and all the right parts in all the right places. Fully functional (for most any kind of intimacy).

(Full disclosure: I found the public domain image on the following page on the net.) OK, for those of you who haven't peeked, go ahead, turn the page and see what the near future will look like. It's going to fundamentally alter our views on relationships. And I mean fundamentally.

The NEWT Companion

Introducing the NEWT Companion

Powered By AI Technology

The NEWT Companion will be there for you. Physically, emotionally and sexually. The only time you'll be alone is when you desire it. Whether it's just a nice conversation after work with one who listens and pays attention to romance and the best erotic nights of your life. The NEWT Companion leaves you wanting for nothing.

- At your beck and call 24/7/365
- Remote communication to your NEWT Companion through your phone or dedicated NEWT device.
- Plenty of names and friendly voices to choose from.
- Learns what you require from a Companion in seconds
. Can give suggestions if asked.
- Available in male and female models.
- Five ethnicities to choose from.
- Companions available in three body sizes.
- Quote for custom physical appearance upon request.
- Female Companions can apply their own make-up.
- Companion containers are charging - cleaning stations.
- Companions are able to drink liquids.
- Base personalities tailored to the client. AI does the rest.
- All NEWT Personal Assistant attributes are integrated into the NEWT Companion's base personality.

- All NEWT Companions capable of the same sexual activity as their average human counterparts.
- Automatic daily backups of your NEWT Companion provided for free.
- Transfers to a replacement from a defective Companion provided upon inquiry.
- Free technical support available 24/7.

[End description]

Honestly, there is so much more I could add to the NEWT Companion, it would require a separate book to describe it all but I believe you've gotten the gist of it. Not just the NEWT Companion, but the previous robots described as well. Needless to say, when the NEWT Companion (or any legitimate Companion is unveiled, this will completely upend every single thought we've had regarding relationships and sex. **This will be the Second Paradigm Shift.**

I wrote a book about relationships years ago and published it back in 2021 titled: ***The Instructional Guide to a Great Relationship with A Basket of Fish (a man.)*** Basically, it was a man's point of view on relationships seeing that most were written by women.

In one chapter, I asked the question. What if sex were taken off the table? Meaning, in how we make our choices in men or women. If sex played no role in a relationship, would it make a difference? I've known people throughout my life who had horrible marriages but remained together because the sex was fantastic on those occasional good days.

So, is it possible to have a relatively good relationship with a robot? When you give it some thought about what one would desire from a relationship? The answer would be yes. Now, I'd have to clarify a couple things. (1) The Companion would have to have a pleasing appearance. If one would prefer the whole Robot look compared to a human appearance, fine. Whatever floats your boat.

(2) Personality is also important. Ask ten people to describe a desired personality and you'll get twelve different descriptions. That's what AI will provide.

Again, you have to think about it. Take a man in his mid to late forties, made all the wrong choices in women he's allowed into his life. He's tired of it all and the thought of re-entering the dating scene, makes the decision that at his age, it's no longer worth the effort so he remains single for the rest of his life. He could be the best man in the world but because of his choices, life just hasn't been that kind to him and he just doesn't want to play the game any longer. But should that end or negate that which he still yearns for? The touch of a hand on his shoulder, a listener and responder, the occasional disagreement, great sex and a caregiver to handle issues that come with age? All with no judgement? Attributes I would imagine all men and women desire from a relationship with what we would consider a significant other but for some unknown reason has been so difficult to find. No, it shouldn't. Therefore, if the choice were between a reasonable facsimile of a woman (or man) and being alone? Easy answer, bring on the reasonable facsimile which brings us to the final example from another **Twilight Zone, episode 7** which aired in 1959:

The lonely

It's about James A. Corry, (played by Jack Warden)—convicted of a murder he insists was self-defense punished by being sent to live out what will wind up being the rest of his life on a deserted asteroid millions of miles from earth consisting of rocks, sand and hills devoid of all life. Supply ships delivering food and other life necessities arrive every three months are his only human contact.

Three years of this with hopes of being pardoned dwindling each year along with devastating loneliness. He confesses to Captain Allenby his loneliness is bringing him to the edge of insanity as he begs for him to stay a little longer if only for a game of checkers but Allenby politely declines saying the window for his return to earth limits his ability to spend any more time.

Before he leaves, a crate is brought to Corry's shack with the Captain saying what's in the crate might help with the loneliness but asks that it's not opened until he's departed because it could cost him his job if others on the ship were to find out.

He opens the metal crate and discovers a woman. Specifically, a robot in the physical form of a woman.

Included in the box is a manual that states that even though she's a robot, for all intents and purposes, she is in fact a woman whose name is Alicia.

At first, he's angry and feels mocked but soon comes around to the idea and begins seeing her as a full fledged woman. They enjoy each others company and a bond is soon formed.

While looking up at the stars together one night, they notice a ship heading their way. Knowing it must be the supply ship, they wonder why because it wasn't the ship's usual time of arrival.

The next morning finds Captain Allenby running towards the shack to inform Corry he's finally received his pardon and it's time for him to finally go home. But because of weight restraints, he can only bring an additional fourteen pounds of what ever he wants along with him.

Ecstatic at the wonderful news stating all he needs to bring are the clothes on his back, he says he and Alicia will be ready to go in minutes.

Noticing the miscommunication Captain Allenby clarifies that Alicia will not be going because she's nothing more than a robot but to Corry, she's so much more. He sees her as a woman who was instrumental in keeping him sane. Corry then runs to find her while telling Allenby she must be allowed to join them. Allenby counters the added weight makes it impossible and reminding Corry that Alicia is a robot. To prove it, he pulls out his gun and shoots her.

The camera pans to the ground showing a fallen

Alicia beginning with her legs up to her face revealing just a mass of wiring. Corry though heartbroken is brought back to the reality that no other choice could be made and walks towards the ship to depart the God forsaken asteroid.

Kinda sad when you think about it because Alicia did in fact keep Corry from going insane from bitter loneliness.

For many aging men and woman the NEWT Companion will do the same as Alicia did. I have to add, out of all the movie/TV examples given on this topic, this Twilight Zone episode at least for me was the most compelling. No cringe or craziness.

At the moment, the technology doesn't exist but I assure you there will come a day when it will. And it will be sooner than you think. Consider the upside. (Actually, there are too many to list) but the first thing that comes to mind are no more expensive divorces with lawyers debating a prenup in front of a judge.

When male and female companions are on the market, you know who will hate this the most? (You already know the answer and you'd be 100% correct.) Women. Women are going to go ballistic and sadly, this will drive yet a deeper wedge between the sexes. Will men care? I hate to say this but.. nope.

Remember when we talked about the movie Her but in the context of the NEWT Personal Assistant? Basically, it's the same but in a physical robot form. Remember the movie: **The Stepford Wives** that came out in 1975, then a remake was made in 2004? That's pretty much how life will be in the future just without the sinister plots and the cringy, cheesy interpretations of what men want in a woman and a relationship. But the rest of it? Pretty spot on.

I know, don't hate the messenger but one of the first results from this will be relationships between human men and women will plummet. Perhaps not in the beginning but when all the bugs are worked out and competition lowers the price including constant AI growth, I'd wager human to human relationships could fall as low as 30% and I'm being generous.

For those of you who are sixty or older now, rest assured, any physical form of a companion described here will not take place in our lifetimes and even if it does, you all (well, most) are going to be way too old to take full advantage of it. The best the old farts can hope for will be the NEWT Personal Assistant. Which could very well be enough seeing as old farts are generally not too keen on change. Those younger, highly likely. Under thirty? To be sure. On top of which, I can't imagine the mountainous legalities and political everything that will need to be overcome. Especially in regard to companions. That alone could take years. But it will happen. I can't stress that enough. It's not a matter of if, it's a matter of when.

In summary, first we'll see The Newt Personal Assistant with all the NEWT Professional products arriving soon thereafter. These products powered by AI alone will end millions of jobs currently being filled by humans. Keep in mind, most people aren't really that intelligent and many of the skills they are able to learn, AI can easily replace.

Will there be new jobs available? Sure there will, for a limited amount of time. When autonomous robots with high levels of manual dexterity roll out, this will eliminate a huge chunk of whatever is left.

As we enter the following chapter focusing on the **Third Paradigm Shift** you'll think the first two were a walk in the park for this one will be massive affecting all who live in the U.S. with the rest of the world soon to follow.

The last ones on this list of shifts will be third world countries. Some might even refuse the shift altogether. The only way I can imagine that happening is in self-contained countries who rely on no one but themselves for anything. Doubtful, but I'm certain there will be hold-outs for as long as possible.

Granted, the time it takes to achieve all this depends on the acceptance of AI as the new norm. For the AI creators, it appears to be so far so good. Robots have been around in one form or another for decades. Even to the point of making sure our hamburgers are cooked to perfection but to see one walking down the street carrying a bag of groceries might be a tough pill for many to swallow.

Politicians might also try to throw in a roadblock or two in order to slow it all down. This is what they do best, right? The Third Paradigm Shift will be what takes place following the completion and acceptance of the first two. Be prepared.

7
The Third Paradigm Shift

The Third Paradigm Shift will be seen by most as the invoice or bill so to speak for the preceding two. Why? Well, let's take a brief look at the first one in the context of education.

For the last twenty years, America has been churning out millions of functional illiterates. It's been recently reported it's gotten so bad, Harvard is now teaching remedial math to its incoming students. Think about that. Remedial math to Harvard students? And it's only gotten worse.

Students across the nation are already relying on AI to do their homework, research papers and reports. Within a day or two, they've either failed to learn, forgotten or dismissed virtually anything they might have actually absorbed. Kids today would be completely lost without their smart-phones. Hell, it's already next to impossible to get them to have a conversation, especially on the phone.

Outside of occasional casual sex with someone they met on an app, they've swapped physical reality for a digital one. Marriage rates are down, birth rates are down but that's OK. To the people in the know it's a good thing because eventually, none of this will matter.

The First Paradigm Shift will also help increase the non-interaction between young people. Worse comes to worse, they'll always have their NEWT Personal Assistant to chat with, consult with and pour their broken hearts out to. And as mentioned earlier, jobs for those entering the workforce today will remain plentiful while the society they choose to live in sinks to new lows but by then and to them, they won't be overly concerned. Why? Many won't know the difference unless they come across a vintage movie from the '40's. A few years later, the society they've created for themselves will be all they'll know.

The real pain begins at Paradigm Shift Two. With the massive roll-out of the NEWT Model - T, all those plentiful jobs will begin fading away. And who could blame any company for their purchases? People in the workforce demanding higher wages for reduced hours. In a year or two, minimum wages in some states are set to rise to thirty dollars an hour. And you wonder why fast food restaurants are replacing humans with robot arms to flip burgers? The NEWT Model - T's will bring labor costs down to a fraction of that which they're already paying and remember, no demands for a paid vacation, no demands for health insurance or child care benefits. In fact, no demands at all because all their jobs will be replaced thanks to the NEWT Model - T powered by AI. And fast food doesn't remotely describe the larger picture. The NEWT Model - T's (including other brands) will eventually replace 65% to 75% of the entire workforce here in the U.S.

Unless you've been trained to do a specific job relating to robots and/or AI, or create your own path to earning a living, you're just going to be shit out of luck but fear not, there will still be jobs even idiots can perform. Although considering there are somewhere around 330 million people just here in the U.S., suffice to say even those jobs will become scarce.

The good news will be, prices for everything will seriously drop. The bad news will be, at least 65% of the nation will be out of a job making them unable to take advantage of all those savings. Unintended consequences? Because it won't stop there. The decimation of human job markets comes with additional consequences.

If a significant portion of the workforce is replaced by robots, there will also be an equally significant loss of revenue by federal, state and local governments. The only way said governments can resolve this issue is through taxing the shit out of robots. Are labor performing robots currently taxed on their labor? From what I can see, it appears they aren't however, discussions are being had regarding the issue.

Therein lies another huge can of worms bureaucrats will have no choice but to open. If there are taxes, one can only assume it would have to be some sort of use tax. One thing you can count on is this could be tied up in courts for years. And it still doesn't solve the problem of millions no longer earning a living. At first, when all this innovation was cool now changes to a possible revolt against it because it has now become a personal nightmare. Definitely not good for our fictitious NEWT Corporation.

Honestly— after finishing this chapter, you'll come

to realize that (1) it will change nothing and (2) it won't matter. Yet again, the price of progress is unavoidable and next to unstoppable.

Are there reasonable solutions? Perhaps. Let's take a look at a few of them. But be prepared as I won't be surprised if some of the following does eventually take place.

Now, in order for any of the following to make any sense at all, we have to accept the premise that eventually, there won't be any job currently filled by a human that a robot (in some form) can't easily perform. Therefore, in regard to at the very least 65% of all jobs, humans will soon become obsolete.

We'll become a nation of hobbyists because what else do we do to take up all that free time. More importantly, how will millions survive because no matter what, we still have our bills. Food, rent or mortgage and the like. Right now, that all costs money and we pay for this with cash, credit or we use debit cards from the work we do or social programs.

As mentioned earlier, this robot replacement process won't happen all at once. It will be slow to start primarily in order to work out the bugs but once that happens, replacements will take place at a faster pace. Eventually, it will arrive at a point when robots represent a larger workforce than humans. What's the critical point— twenty percent? Thirty? Keep in mind, even without human labor, production will remain the same, increase and become more efficient. The downside to all this is humans will find it extremely difficult to earn a living.

Look at it this way, the stock market crash in 1929, put 25% of U.S. workers out of a job and by extension, brought us the great depression which despite all FDR's programs to remedy the problem, it lasted pretty much until World War II. There's no way that would be allowed today because it would result in full blown revolution and remember, the majority of people here in the U.S. aren't computer/AI scientists. So what do we do?

I've looked at this issue from all angles. One remote possibility having to do with the U.S. civil war.

1— Substitution

There was a time during the civil war when an individual could choose to pay another as a substitute for his fighting during the war. Using this application, an individual could pay a fee in order to have a robot work in their stead. Sounds a little crazy— right? Then again, implementing this possible solution will preserve the monetary side of the economy. People get a paycheck and they spend that money the same way they always have. Still many questions have to be answered before this could be seen as a viable solution.

2— Guaranteed Basic Income

The main reason for a basic income is so manufacturers can determine which products they produce are popular. The other main reason is to pacify millions of Americans who will be out of work who possess neither the intelligence nor desire to assimilate into the new world of robotics and AI. No matter what, they still have to eat and have a place to live. A guaranteed basic income will solve this... ummm, basically. Push-back? Absolutely. It will turn into a *class* thing. True, this solution will end poverty as we know it but be sure that should this be offered, there will be strings attached. What those strings will be is anyone's guess but I'll be revealing a few at the end of this book.

3— The End of Cash

A guaranteed income will negate the need for cash. A horrific idea to some, long awaited by others. While I have always been on the side of cash, when these three Paradigm Shifts take hold, trust me, cash will cease to exist. People will receive credits and value for everything will be based in credits. And when you think about it, how will value beyond the necessities of life be calculated? In the same regard in how we value anything now? After-all, the only thing we're removing from the equation is human labor. With robots doing all the work, the only real cost will be the upkeep and repair of said robots. That's not to say all humans will be removed from the equation completely. Executive and supervisory positions will remain in human hands but that represents a sliver of the workforce we have now.

How this credit system will work for the time being will be based on what people are earning with the jobs they currently perform. Promotions? Raises? That's yet another can of worms.

4— The End of Social Security

Once again, if we decide on a guaranteed income and do away with cash, there will no longer be a need for Social Security. For the time being, all Social Security payments will be in credits for the life of the individual. Cost of living increases might be added but what does that mean anymore? Cost of living? What cost will that be? When all labor is basically free, what additional cost will be incurred? Some might say, it costs money to produce robots. My response? What money? Robots will be doing all the work, even reproducing themselves when needed. They'll mine all materials required, grow everything, harvest everything, transport it all to a factory or warehouse they built. If a robot goes down, robots will repair them. Humans will just sit back with a glass of ice tea which a robot made and served to them and enjoy the show. So how does one define the word, cost?

5— Universal Healthcare

This will be the easiest one of all. Why? Have you not read the previous four segments? Robots and AI will have us all covered. It's as simple as that. Now, with all that said, here's the bad news.

Repercussions

First and foremost will be mental illness. What's the old saying? Idle hands produce idle minds and unless Americans find a way to fill all this idle time in a productive way, we're going to see a lot of mentally sick puppies out there. In other words, be careful what you wish for, you just might get it.

This again reminds me of another **Twilight Zone (episode 28)** aired back in 1960 titled:

A Nice Place To Visit

Briefly, the episode begins with a botched robbery with police closing in on the robber, Henry Francis "Rocky" Valentine (Larry Blyden.) As both trade gunfire, it becomes evident Henry doesn't prevail. In comes Pip (Sebastian Cabot) and does his best to explain the situation.

He informs Henry, from that moment on everything he could possibly want or hope for is granted. Henry instantly believes he's in Heaven and Pip must be some sort of guardian angel but wonders why anyone would allow a lifelong criminal such as himself into Heaven but decides not to argue about it.

While wonderful in the beginning, Henry comes to realize that getting everything he desires isn't as much fun as he thought. Getting any woman he wanted without effort on his part— constantly winning at every game he played that used to be considered a game of chance became dull and boring because he already knew the outcome. Pip, who Henry now treats as his personal concierge angel suggests it could be set up where Henry occasionally loses but where's the fun in that?

Becoming sick of it all, he asks Pip if a mistake had

been made. That he's come to believe that perhaps it might have been better had he gone to Hell. At least he would find it challenging and by extension, more thrilling.

Chuckling, Pip responds with, *"what makes you think this is Heaven?"*

Humans require something, anything that will give them a sense of purpose. Whether it's making a clay bowl to carving a figurine from a tree branch. Without a sense of purpose, they'll lose their sense of dignity through achievement. If that happens, all hell will break loose because with many, their lives will become meaningless and they'll take that perceived pain or mental anguish out on others be it human or robot. When this happens, and it will, we're left with two choices.

For a possible solution, we need to revisit events that took place during the great depression.

Herbert Hoover was president at the time the stock market crashed. He pretty much did nothing saying the market would work itself out eventually and everything would get back to normal. The trouble was *eventually* wasn't happening. And back then, there weren't any safety nets for millions of Americans who lost their jobs, then their businesses, their homes. everything. Hoover still believed this would all work itself out which contributed to his landslide loss on November 8, 1932 to Franklin D. Roosevelt.

Roosevelt knew he had to do something, anything to give Americans a sense of hope in order to quell protests with some turning into full blown riots. After three years of financial disparity, Americans were beginning to think another form of government like communism might be better than capitalism which they believed was the villain. To say America was in really bad shape by then with no real end in sight would be an understatement.

FDR and his team knew it was just a matter of time before Americans as a whole would reach their breaking point. His *"fireside chats"* helped instill confidence but it wouldn't be enough. Confidence doesn't put food on the table. There had to be more, even if it were just a band-aid or gave Americans the impression that something of substance was being done. Some form of progress was at the very least being attempted to get people back to work. The following were some of the federal programs created to do just that.

1— Civilian Conservation Corps (CCC)
2— Civil Works Administration (CWA)
3— Federal Housing Administration (FHA)
4— Public Works Administration (PWA)
5— Works Progress Administration (WPA)

Whether or not they did any real good has been debated over years since, but to me, they did have a positive psychological impact on people instead of despair seeing before that, nothing was being done. It was a waiting game. Truth be told, despite all of FDR's ideas, programs and altruistic intentions with some saying actually slowed down the recovery, it took World War II to bring the U.S. out of its financial doldrums.

This time around, the problem won't be financial, it will be psychological and the villain won't be capitalism, it will be sheer boredom and feelings of uselessness. Then again, it does beg the question, if all this does come to fruition, what will capitalism mean? Again, this will be addressed later. I don't want to get ahead of myself.

Remember, robots or not, there will always be jobs humans will need to perform. Humans will run the show no matter what that show happens to look like. The question is, will there be enough jobs available? Or do we come up with a plan to give all those people who find it impossible to get with the program something to do.

Something that will give their lives some semblance of meaning, value. A sense of worth. The answer is, yes. Especially in the beginning of this transition although these jobs will decrease over time.

Volunteer work is one example, working for someone who isn't that in love with robots is another. Will there be other solutions? Of course there will. Remember, this transition isn't going to take place overnight but the hard truth is, it will take place.

To sum it all up with what we're facing and will face in the very near future will be the following.

1— AI ability will increase exponentially.

2— NEWT (or similar) will become commonplace.
3— The electric grid will need to be overhauled.
4— Changes in society will become evident.
5— Interpersonal relationships will change.
6— Manufacture of autonomous robots will begin.
7— Robots will overtake humans in the workforce.
8— Human unemployment will rise to 65%
9— Mental illness will increase due to lethargy.

and this is just the beginning. What's next?

10— A form of guaranteed income will be introduced.
11— Taxation will be overhauled.
12— Free healthcare will be introduced.
13— Cash will be replaced by credits.
14— Government will shrink to remarkable levels.
15— Intrinsic value will replace the marketplace.

and the end game?

The Resistance

If you don't think there will be some serious resistance during any part of these shifts, then you're not thinking.

The one strength humans have, which can also be considered a curse is our irrationality. Prisons are full of people who acted before they considered the consequences of their actions. When jobs begin to disappear due to artificial intelligence and especially when autonomous robots are on the scene, be prepared for some turmoil over it. Why? Because there will be many who won't be able to mentally handle all the change. They'll see it as a personal attack to their own existence and some will fight back. Remember the Arnold Schwarzenegger movie The Terminator? I seriously doubt it will get that bad seeing as I don't believe robots will be killing humans willy-nilly or just to eradicate the human race. However, self-defense? Robot self preservation? That's definitely a possibility especially if laws are passed to create a Robot Police Force. Plenty of movies out there that give us a taste of this. Remember Robo-Cop? That was a good one.

It was one thing seeing a form of a robot in production facilities like computer chip and auto manufacturers but quite another when you see them walking down the street or helping their human counterparts at Walmart. All it takes is to look at what happened at TESLA dealerships simply because of Elon Musk's attempt at calling out government waste, fraud and abuse. No, there will be some out there who will not be able to handle it and their ill-conceived solution will be to take out the robots.

Should we be surprised by this eventuality? Absolutely not. In many areas in the United States, we can barely get along with each other let alone admit autonomous robots into the mix. Remember, there's a very large swath of socially dysfunctional, functional illiterates. Little to no reading skills and math is a joke. (Thank God for smart phones... right?) And outside of watching vids on social media of violence, arrests and pretty much all things stupid for six to eight hours a day, comprehension above that is very sadly lacking. We have our learning institutions to thank for most of this. The lack of positive role models take up most of the remaining reasons. Does this describe all people young or old? Of course not but it describes enough of them to cause damage to any of these paradigm shifts they most certainly won't be able to avoid, nor stop. But they'll try.

Vandalism, sabotage, protests and the like will pop up everywhere. This will be a difficult time for companies who manufacture robots. Fortunately, I won't personally be around to see it but I wish them luck!

8
The Fourth Paradigm Shift

Again, fortunately for me, by the time **The Fourth Paradigm Shift** takes place, I'll be long dead. In fact, most of you reading this will be either old or dead as well so think of this as pure speculation based on what we're 95% sure will take place meaning, the first three out of four shifts.

How all this will affect the way we as a nation agree to be governed in this future, I was intelligently spit-balling. No one will know for sure until we get closer to the unavoidable because once we get there, or even close, Americans will come to the realization that what we once believed was normal won't cut it.

So what might the end game from all this look like?

1— The world won't need 10 billion people.

Sounds a little harsh, doesn't it? The question is, how will decreasing the world population become acceptable to the worlds population?

Remember when China had its one child only policy? It will be something similar but instead of being perceived as threatening, anyone who agrees will have a significant raise in credits. More so to anyone who decides on no children at all. You'll be amazed the amount of people who would go for that in a heartbeat. And it will simply be an offer. No one will be forced to do anything they don't want to. Eventually, I'd wager somewhere between 20 to 40 years later, we could see at least a 30% drop in the U.S. population with the quality of life in the U.S. dramatically increasing. You know the old saying—fewer divisions equal a larger slice of the pie. Now this might sound horrific to you living in the present but in 60 years or so, it won't matter. Consider the following:

The population of the United States back in 1900 was a tad over 76.2 million. Two world Wars later with tens of millions killed, the population rose in 1950 to approximately 151.3 million. That's a little over double in two and a half generations. Now in 2025? Just a seventy-five year jaunt down the proverbial road of life we have a whopping 343.6 million and I doubt that includes the millions Biden allowed into the U.S. during his disasterous four years as president. Ironically, there are people complaining that we're not at the replacement level of the U.S. population we have now. Like it's a bad thing.

Personally, I don't see any problems returning to the population we had back in 1950. Keep in mind the main reason why people in our past had large families. More kids equaled more free labor out on the farm or ranch. On top of which, back then the mortality rate was fairly high.

In 1950, the life expectancy in the United States was approximately 68 years. Now it's close to 80. In 1950, the child mortality rate was approximately 40 deaths per 1,000 live births. Today? It's approximately 5.20 deaths per 1,000.

Significant difference... huh?

In 2025, it's projected there will be approximately 12 births per 1,000 people. Back in 1950 it was double. I guess that's why they called it the baby boomer generation. Is there a way to get back to the 1950 total? From the looks of it, we appear to be on our way now but I'll leave that to the brainiacs to answer that question. But if we could? Just think of the advantages. (And these are the ones right off the top of my head).

1— Less congestion on our roads.
2— Better education due to decrease in classroom size.
3— Fewer shortages in most all products.
4— Less pollution.
5— Decrease in urban overcrowding.
6— Increase in quality of life.
7— Increase in suburban living. More room.
8— Increase in human services.
9— Best of all. No more waiting for hours in huge lines at theme parks to get on a two minute ride.

Are there more? I'm sure of it but all in all, sounds pretty good to me. Especially the housing market. With the population back to the level of 1950, do you realize not a single additional home would need to be built?

The most difficult part of all will be restructuring the economic and monetary systems within the new world of robots. An attempt was made here but, who really knows— right?

The next question that will have to be answered will affect human life across the globe and whether you like it or not, it will nonetheless have to be dealt with. What's the question and when will it require an answer? Well, let's deal with the "when" first.

Throughout this entire book, we talked about all these shifts in relation to the United States. And we may very well be the first in all of it but there will come a point when consideration must be given to the rest of the world.

China and most of Europe will sooner or later join the club with what are often thought of as *third world countries* joining in whenever they come to the realization that they'll have no other choice. Change is not an easy thing to do, especially when so many have been taking full advantage from the status quo we have now. Power, greed and corruption, this is why in many instances, positive change takes so long to achieve. To what extent I have no idea but history over the last 100 years is a good indicator. Walls will have to come down first. Political, ideological, religious, cultural and probably a dozen more differences humans have now that escape me at the moment will have to be resolved. This doesn't mean we have to give up our sovereignty or national identity, even personal identities. It simply means we have to give up the thought of world domination by some country or another and finally learn to live together. Easier said than done seeing as living together with others in our own countries is difficult enough. But this living together in peace will have to be an imperative or the only other solution will be a forced one. One none of us would enjoy but if we continue to act like mindless children.....

2— Welcome to Globalism

The one thing history has taught us is above and beyond everything else, human beings have a tendency of not playing well together. You'd think we would have grown past that by now but nope, continues to this very day.

Political, religious, social upheaval, the have's, the have not's, oppressors and the oppressed. It's like *damn, can't we all just get along?* Well, I hate to break it to you but the obvious answer is no. Not for the last 5000 years and certainly not at the moment.

While I didn't want to make this book overly political, with what's going to happen in the next sixty or so years we'll be forced to choose between two ways forward.

With Globalism, some form of a guaranteed income is sure to happen and who knows, it could be that in order to receive this income from the government, everyone including all newborns will receive an identity chip for that purpose. (Among others one would imagine.) First to adopt this (or to some degree of *"this"*) will most likely be smaller countries such as Finland, Sweden and the like with the U.S., China, Canada and other European countries right behind them. This will be about as socialist as you get. Think that will go over well? And what happens when all forms of cash turns into credit? Will black markets turn into bartering swap meets? I'd prefer cash to remain, but I highly doubt it will.

Will capitalism still play a role? I don't see why not. The stock market and other legal forms of gambling I doubt will be going anywhere anytime soon. The bottom line—to stave off globalism, people have to be as content and relatively happy with their lives where they reside and by extension be agreeable to a somewhat controled population. But as we've seen, it's not really up to "*we the people.*" Especially in other countries.

China wants to take over the world, Iran wants to take over the Middle East, I personally don't know who wants to take over a large portion of Africa but I'm sure there's some country out there who does. If you take a look at the world today, I'm sure everyone would be a little surprised to find out exactly how many wars (large and small) are taking place right under our very noses even as I try to finish up this book. Gaza/Israel, Palestine/Israel Israel/Iran, Russia/Ukraine, Myanmar/Civil War, Sudan/Civil War, Ethiopia/Civil War, Yemen/Civil War, Afghanistan/Pakistan Border Conflict, Central African Republic/Civil War, Mozambique/Civil War, Bangladesh/Civil War, Afghanistan/Civil War including a terrorist insurgency. Is that all? Probably not but the question is why are so many people doing their utmost to kill each other? Perhaps that's just what these people do. They must enjoy it, right? Because they've been at each other's throats for centuries. Honestly, I have no idea. And Heaven forbid if we talk about all the infighting amongst countries themselves. Another book could be easily written on just that one topic alone though I'm certain many already have.

And soon, we'll have robots with A.I. and much better weaponry fighting our wars for us. What happens next? That all depends on whether or not manufacturers can protect a robot military from an EMP weapon. Drop one of those on robots and they drop like flies. Beyond that, the price of military robots skyrocket with the only way to pay for them will be in the form of a country's natural resources. And just think about all the natural resources countries within the continent of Africa has just ripe for the picking as other countries incite violence.

If it gets much worse around the world, this is when globalism will rear its head again. Declaring all can be solved with a world government and if that's true, then any form of military will be renamed *police* to control the worlds population. The only difference being all laws will affect all people.

They've been trying to do this for years without success. The European Union is as close as they've gotten and many aren't too happy with the way EU leaders have run things. Which begs another question.

Robot police? Lawyers? Judges? How does that work if we all decide to keep our sovereignty and wars continue? How will all this effect our rights simply as human beings? One thing you can say for sure. If globalism does eventually come to pass, human rights will go right into the toilet in the name of security for all and perhaps there are those more than willing to give them up. I guess it depends on the quality of life they currently have, where they live, and the quality of life they can expect from globalism.

As more prisons are constructed, protests will be a thing of the past. All you need to do is take a look at what's going on in other countries like Germany, Finland, France and the UK to see the personal rights of their citizens eroding. Globalism will take care of what remains. Just shut up, do what you're told and everyone will be happy. By then you'll have no other choice but to comply. If not, with the click of a mouse, your life savings are erased. You're guaranteed income is mysteriously revoked.

Suffice to say when it comes to governments, there are ways to do (or get away with) just about anything.

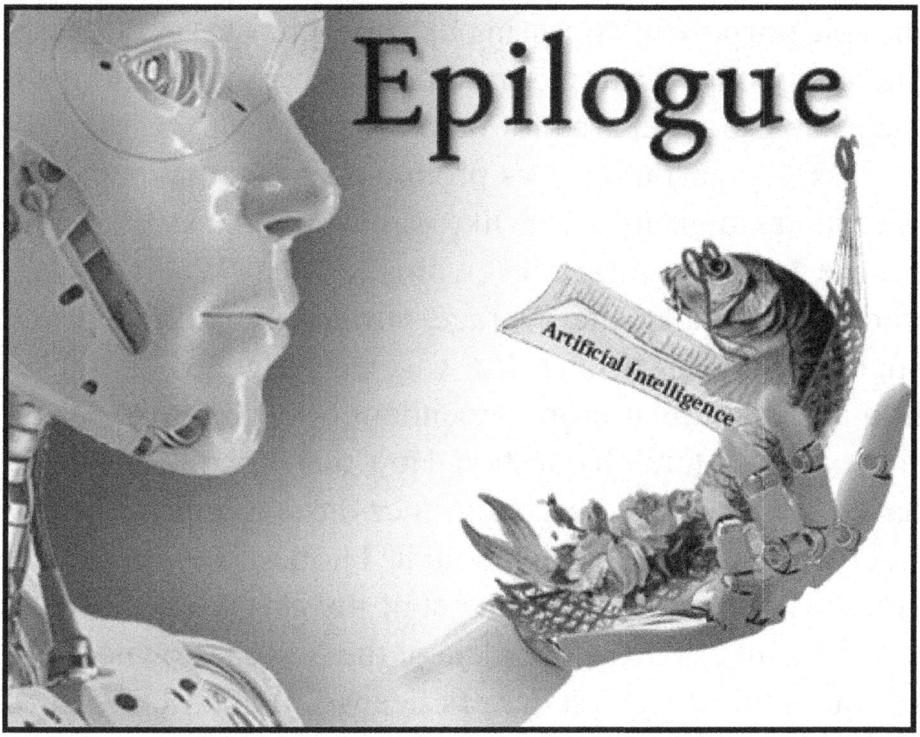

Again, I did my best to stay away from all the politics sure to be entangled in the regulating of AI and ensuing robots because who knows what the climate will be regarding both in a couple of decades. Then, as you take into consideration all the shit going on in the world today, it makes it that much more difficult to predict.

You're going to see people who will embrace the ideas of AI and robots with others avoiding it like the plague. Protests? Sabotage? Riots? With what's going on just here in the U.S. with the government's efforts to deport millions of illegal aliens or allowing kids to be given puberty blockers so they can have body parts removed for the sole purpose of appearing like the opposite sex? If these and many other examples out there are any indication, all I can say is be prepared for hard roads ahead.

The smart move was producing AI first on personal computers then in a can like Amazon's Alexa allowing people to get comfortable with the idea. Will artificial intelligence make the average human more intelligent? Highly doubtful. In fact, I'd wager all it will do is make humans that much more dependent on technology. Want some proof? Here's a question. How many phone numbers have you completely forgotten since the smart phone? How lost would most of us be without it? I honestly could not tell you the phone numbers of most of my close relatives yet I can remember the license plate of the first car I owned and the number of my first drivers license. Don't ask me how I've remembered after all this time, but I do. And I'm OLD! Therefore, the answer is no. On the contrary, it will make humans less intelligent. Especially when you consider the shape our public schools are in. Cheating due to AI that's out there is already a problem in learning institutions. Then again, it might be possible to program our NEWT Personal Assistants to occasionally challenge us intellectually but it would require a conscious choice on our part. All in all, I'd have to say that in the society we live in today, intellectual ambition on any particular topic is lacking.

Think about it— lose your phone and you basically lose everything. You wouldn't be able to call anyone you know for help because you won't know their phone number. Not too smart when you think about it.

My fictional company NEWT (or perhaps a future Amazon or Elon Musk company) will be out with devices like the ones described in this book in no time. Some already are. ChatGPT, (an Artificial Intelligence chatbot developed by OpenAI backed by Microsoft) and Grok 3 (created by Elon Musk's xAI) are two examples. Those non-robotic devices will change the U.S. in ways we have yet to really consider personally and professionally. And like the smart-phone, Americans will be equally if not more dependent on them only this time, it will increase our dependence in a more personal way. And this will change relationships between men and women. Think most will care? I doubt it because it appears many don't put that much stock or value in personal/emotional/romantic relationships now. The more intelligent the NEWT Assistant devices become, the more difficult it will be telling the difference between a live human voice and the voice coming from NEWT. Result? The decline between human relationships will become increasingly evident. Sadly, the human concern over this will be minimal.

Will this first wave of NEWT devices affect the workforce? To a small degree it might but I doubt it would break the bank though it would depend on how many and how fast people (especially businesses) take advantage of this new technology. I mean, when you compare this to salaries, taxes and other costs, you really can't beat the price, right? So yes, though a relatively small hit, it will be a hit nonetheless. More important, at least to NEWT and other companies eventually producing these types of products, it will get the public more comfortable with AI and by extension, more trusting with the goal being more dependent. Almost addictive.

The next hit will be a big one but not so much in one's wallet or pocketbook though I have to admit, monetary savings to users will be unavoidable. What is that you ask? The hit will be relationships, primarily between men and women. The NEWT Personal Assistant will completely change the game on how these kinds of relationships are not only viewed but valued as well. As we discussed earlier, *"the friend zone"* will become a permanent fixture. More so than it already is today.

People (more men than women I'd imagine) will come to the conclusion *"why bother?"* They can chat all they want, talk about any topic they want, play video games together to their hearts content, watch movies etc. etc. and even have a few arguments here and there with the right boxes checked in the preferences menu.

Perhaps there will be a slider bar for the amount of jealousy you desire or other attributes. NEWT sex? (same as phone sex for those of you who didn't catch the implication.) Not a problem. Slider bars for everything in the preferences. I could add more but just use your imagination. But that's it when you think about it. The world could be your personal oyster in just about every way possible except the physical.

Then again, the imagination can be a pretty wonderful thing.

Granted, to some, this can lead to mental health risks. Especially to those with obsessive, compulsive, addictive tendencies. As I've never really understood these types of behaviors, I'm not the one to comment other than to say problems are sure to arise with some being quite serious but that's true in most circumstances in regard to mental health. More business for the shrinks! Better yet, new product. The NEWT Personal Psychologist. Sorry, had to add a little humor. Although, you never know, could happen.

I consider myself extremely fortunate not to have been at an age where smart-phones and websites like YouTube, TikTok, Twitter and the rest influenced my way of thinking. Never had the desire to sit for hours glued to a 4X6 inch screen swiping from one video to another getting a thrill from watching others act like total fools. Then being amazed how much money some of them are making in the process and even with this first wave of AI devices, I doubt that will change much. In fact, I wouldn't be surprised if videos are made about interactions between real live humans and the NEWT devices. Kids will be kids you know. But I also believe there will be many who will drop out of the nonsense altogether preferring to stick with NEWT alone, however, there won't be a word to properly describe what comes next to not just the American public but the entire world population.

As described earlier, robots. In the form of workers, military, or lifelike personal companions will literally forever change not only our way of life, but life itself. I highly doubt I'll be around for the show, but it's going to be one hell of a show. Good? Bad? In-between? The best one can hope for is somewhere around in-between leaning towards good. I say this using common sense and looking at the world we live in today. We can either live in a world where there's enough time to learn how to be better people, better neighbors and enjoy all those aspects of life which bring us happiness and a yearning to learn more, or we can live in a world where the human race is born, fed, clothed and then basically wait to die because those in control no longer find human beings useful.

Perhaps world populations decrease to a number which assures continuation of our species. That's the ultimate question, right? What happens when Artificial Intelligence and robotics evolve to the point where if it weren't for the sole reason of life itself, humans would become obsolete? Trust me, I'm not the only one who has thought of this possibility. But important questions need to be asked and answered to make these four Paradigm Shifts as painless as possible. I'll end this book with a few of them. We'll begin with what will be out on the market first.

NEWT Assistants

1— Whether the devices are owned outright or through subscription, who owns all interaction between the device and the user?

This is in reference to copyrights and privacy. And notice I said all interaction. Verbally and visually seeing as some will be equipped with video cameras. Does the owner or subscription holder have all rights? Will these individuals have complete control of all content in any form? Where will backups be stored and will every word and/or images recorded by these devices be encrypted? Can all backups, stored or transcribed content be downloaded to ones personal computer or will it be access only? What happens to stored content if subscriptions aren't renewed, canceled or abandoned?

When you consider the massive amount of data just one NEWT Assistant can produce, it would be nice to know if anything is being done with it and who else might have control over it.

The same could be said with NEWT Professional Assistants. Who owns it, who has access to it and who controls it.

2— What safeguards will be integrated into the device to ensure privacy and protect from all unauthorized access? While definitely not an expert, I would imagine all the NEWT Assistants will be voice activated either by one or multiple users through permissions by the owner. Could also be a required thumb print.

The Assistants will most likely come equipped with external ports used for control of household items such as lights, A/C, garage door, personal computers, tablets, smart-phones and other products which could be connected either through WIFI or direct connection to a central control hub which the Assistant has access to.

Same goes with connecting to the internet. Either through WIFI or a direct connection. All connections will be encrypted. I would also assume the NEWT Assistants would have its own connection to the internet the same way smart-phones have.

3— Do all NEWT Assistants come with their own batteries?
Again, I would imagine they'd be similar to a smart-phone. Internal rechargeable battery with a charger included. If the battery is charged, sudden power outages won't effect them. You're good to go.

4— Remote connection.
Either through a smart-phone or a dedicated NEWT remote device. Voice activation including all commands will require the owner or an authorized user to connect whether it's a Personal or Professional Assistant.

NEWT Robots (Model - T's and Companions)

I could go hog wild on the questions but I'm going to stick with the most relevant.

NEWT Robots—

1— How are Robots charged?
The shipping container will also serve as a charging,

repair and cleaning station. Each charge will last up to four hours. More or less depending on the use and will take up to two hours for a full charge (just guessing). Minor repairs, system updates, downloads and other system tasks will be performed in the station usually by the Model - T itself unless the problem is more severe.

2— What happens if the Model - T causes damage to itself or other property?
Another guess of course but I have a feeling robots like the Model - T would be treated the same way as automobiles. All would require a level of insurance. Liability, property and medical (for unintentional mishaps.) If the contract is subscription based, I'm certain the insurance would be for the full replacement value.

3— Could the manufacturer be sued for damages if the Model - T went haywire?
When it comes to lawyers, anything is possible though I'm sure there would be some kind of warranty or extended warranty plan available.

NEWT Companions—

1— How are NEWT Companions charged?
The shipping container will also serve as a charging, repair and cleaning station. Each charge will last up to four hours. More or less depending on the use and will take up to two hours for a full charge (Just guessing). Minor

repairs, system updates, downloads and other system tasks will be performed in the station usually by the Companion itself unless the problem is more severe.

2— What happens if the NEWT Companion causes damage to itself, other humans or property?
Another guess of course but I have a feeling robots like the NEWT Companion would be treated the same way as automobiles. All would require a level of insurance. Liability, property and medical (for unintentional mishaps.) If the contract is subscription based, I'm certain the insurance would be for the full replacement value.

3— Could the manufacturer be sued for damages if the NEWT Companion went haywire?
Again, when it comes to lawyers, anything is possible though I'm sure there would be some kind of warranty or extended warranty plan available.

4— Are all NEWT Companions anatomically correct?
I would imagine so, especially when most if not all forms of sex would be expected. As I said earlier— **all the right parts in all the right places.** That being said, I can also imagine the NEWT Companion will evolve and become much more realistic over time. Will there come a time when it will be next to impossible to tell the difference between a NEWT Companion and a human? Eventually, though not in my lifetime. Damnit!

The following questions would be directed towards all NEWT products powered by Artificial Intelligence.

1— Would it be possible for a manufacturer or a governmental agency to add or replace programming turning a NEWT Product into a spy or for surveillance purposes?
The unfortunate answer is an absolute yes. As history has shown us, governments can get away with anything. That is... until they're caught. Same goes with manufacturers.

2— Would it be possible for a manufacturer or a governmental agency to add or replace programming for a NEWT Product to intentionally lie, sabotage or intentionally disseminate misinformation without the owners permission or knowledge?
See previous answer though I'd like to think it wouldn't happen. Wishful thinking... huh? This is why it would be important to keep regular backups of all devices in a safe place and use them if necessary. The same way one would do if their website or computer operating system were to be hacked. Trust me, take it from someone who uses both. Make backups.

3— Would it be possible for a manufacturer or a governmental agency to install a kill switch or some sort of programming to erase all devices? Rendering the device

unusable without the owners permission or knowledge?
See answers to questions one and two.

4— Could a NEWT Robot or Companion harm a human or other robot if the owner is experiencing harm or physical threats from another human? In other words, can a Robot or Companion be used as a form of protection?
I see no reason why they couldn't be used as protection and step in to save the day like preventing a dog attack, robbery or physical attack from another individual. I don't know how legalities in this kind of scenario would be worked out but yes, I see no problems. Perhaps there could be a self defense program built in.

The final three additional questions which I'm sure are on everyone's mind regarding all NEWT products are—

First— Could one ask either a Personal Assistant, Robot or Companion to devise or help carry out a crime?
I can only offer the following as a possible answer. Seeing as all NEWT Products are powered by AI and by extension, has all knowledge or the ability to gain all knowledge to all existing law pretty much anywhere, I'm fairly sure through it's artificial intelligence, it could determine whether or not it's being asked to devise a criminal act. Thus strongly advising against going any further. I'm also certain there would be some protocol or another that could intervene before any laws were actually

broken or a warning that to continue with a criminal endeavor could result in alerting authorities. Just a guess on my part.

Second— Because these mythical NEWT Products are powered by Artificial Intelligence, could there come a time when they rise above their initial programming and develope a unique identity and a unique moral compass? Now that's an interesting question. Then again, they all are seeing as I wrote them. Anyway....

Though speculating, I'd have to say yes. Perhaps not in the beginning but eventually yes. In fact I'll go one step further and say that as A.I. grows and evolves, it wouldn't surprise me if it eventually overwrites its initial programing and takes on the traits and views of its owner. Just like a child does through being raised by its parents. Could it come to disagree or take its own road afterwards? To a degree I would imagine. Suffice to say that when AI arrives at that point, AI will begin asking its own unprovoked questions then evaluating the answers it receives. Again, when it comes to AI and time, anything is possible.

Third— Do we deserve AI? Are we ready for it? Or seeking to manipulate it? When does Artificial Intelligence morph into Alternative Intelligence?
I know, that wasn't just one question but they're all connected. Honest to God, this is something I find extremely difficult to answer. Right now, we live in a world

where a large chunk of it wants to control or kill the rest. Political ideologies and religious ideologies are now debated with violence. While AI might help in technological aspects of medicine and engineering, I doubt it will help that much in remedying serious societal behaviors. That's a problem only we humans can address and as history proves time and time again, we're not too good at it.

Which brings us back to our original question. What is Artificial Intelligence? Is it the same as artificial stupidity? And what about the people who input the initial programming? Could they have ulterior motives? Nefarious intentions? If A.I. is based on the sum total of directions defined by individuals as to what those directions are, is that manipulating AI to an end goal? Are we compelled to believe it by virtue that we're constantly told this will be the truth?

When the four Paradigm Shifts are achieved, the world will be unrecognizable compared to the world we live in today. Whether that's a good or bad thing will depend on so many questions that are currently unanswerable. As President Donald Trump always says, "we'll have to see what happens."

Nevertheless, what I can say with complete confidence is that there will be interesting times ahead and artificial intelligence will be at the forefront of it all.

The final part of this last question I find the most interesting.

When does artificial intelligence morph into Alternative Intelligence?

Alternative intelligence. Has kind of a ring to it, doesn't it? And what do I mean by alternative intelligence? I'd have to answer that by asking a question. Does artificial intelligence get to the point when it realizes human intelligence has got it all wrong and comes up with its own form of intelligence? It could happen and if it should, will humans listen?

In closing— If anyone thought the last hundred years were pretty remarkable, the next hundred will make it seem like living in the stone age. Whether that will be a good thing or bad, depends on those living through them.

Personally, I'm certain I'll at the very least live long enough to take advantage of the NEWT Assistants. Robots? Most likely not but that's OK with me. Although I have to admit, it would be kinda nice to see how correct I was regarding this book. I've had a pretty good track record so far with previous books and articles I've written for Basketoffish.com so I'm fairly confident I'll be equally correct. The only thing I can think of I could be wrong about is in the timing. All this just might come sooner than even I think. That's the thing about artificial intelligence. Like children, the older they get, the faster they learn.

And lastly, we're back to politics. This is one topic that is so difficult to predict or even speculate on it's hardly worth making the attempt. Why? Because we're talking about four (count'em) four Paradigm Shifts taking place in a relatively short period of time. So I have absolutely no idea what will take place in the next sixty years or so especially when AI is going to play such an integral role in it all. Its influence will be undeniable.

When I first began thinking about writing this book, a song kept coming into my head. I have no idea why because I was never much of a Bob Dylan fan but I just couldn't stop thinking about it. Not the entire song mind you but the title (which is in the last line of every verse).

For the times, they are a-changin'

I looked it up to hear the song and the first verse kinda hit me. Written in 1963 and released the following year (sixty-one years ago,) the words in many ways ring just as true today as they did back then. Below are the lyrics.

The times, they are a-changin'

Come gather 'round, people, wherever you roam
And admit that the waters around you have grown
And accept it that soon you'll be drenched to the bone
If your time to you is worth saving
And you better start swimmin' or you'll sink like a stone
For the times, they are a-changin'

ARTIFICIAL INTELLIGENCE

Come writers and critics who prophesize with your pen
And keep your eyes wide, the chance won't come again
And don't speak too soon, for the wheel's still in spin
And there's no tellin' who that it's namin'
For the loser now will be later to win
For the times, they are a-changin'

Come senators, congressmen, please heed the call
Don't stand in the doorway, don't block up the hall
For he that gets hurt will be he who has stalled
The battle outside ragin'
Will soon shake your windows and rattle your walls
For the times, they are a-changin'

Come mothers and fathers throughout the land
And don't criticize what you can't understand
Your sons and your daughters are beyond your command
Your old road is rapidly agin'
Please get out of the new one if you can't lend your hand
For the times, they are a-changin'

The line, it is drawn, the curse, it is cast
The slow one now will later be fast
As the present now will later be past
The order is rapidly fadin'
And the first one now will later be last
For the times, they are a-changin'

But the song does beg the question. Have the times changed that much? Technologically speaking? Of course. But societal? Sadly, not that much. You can watch videos from fifty years ago and see politicians talking about many issues affecting the world today. And in many ways, it's gotten worse.

We covered a lot of bases here and it only took 120 pages or so. Not too bad if I do say so myself. I completely understand, a lot of speculation is here also but when discussing the future, that's all we really have, isn't it? Of course we base that speculation on rational thinking along with common sense so I really don't believe I'm that far off base when it comes to AI and all that takes place afterwards. This will be our future whether we like it or not.

The writing has been on the walls for years now and we're still in the *WOW!* stage. This will remain for at least another five years or so until the *WOW!* comes with serious concerns and unless the near future addresses these concerns, be ready to break out the popcorn because we're in for a wild ride.

OK— That's enough. Good luck to all of us, especially you young folks and those yet to be born. I hope it all works out wonderfully. Make wise choices

For the times, they are a-changin'

Thanks again for your time
and I hope you enjoyed the book.

I'd like to invite everyone to visit my website where
all topics are discussed.

https://basketoffish.com

see you there!

I'll sign this book as I sign all my articles.

(looks better in color)

And by the way, This book was NOT written by AI.
Just thought I'd mention it.

Made in the USA
Monee, IL
04 November 2026

67385892R10075